UNDER
THE
RADAR

A TAKING RISKS NOVEL

Sandy Parks

SANDY PARKS

TAP
TRUE AIRSPEED PRESS

Under the Radar is a work of fiction. Names, characters, places, and incidents are products of the author's imagination or are used fictitiously and are not to be construed as real. Any resemblance to actual locales, organizations, events, or persons, living or dead, is entirely coincidental.

Cover art by Earthly Charms
Formatting by Author E.M.S.

Other books by Sandy Parks

Romantic Thrillers

Hawker Incorporated Series:
REPOSSESSED
OUTFOXED

Taking Risks Series:
UNDER THE RADAR
OFF THE CHART

This book is dedicated to the amazing women pilots of the Spaceport 99s who have encouraged my writing endeavors. Your intelligence and enthusiasm for aviation has never ceased to amaze me. But it's your grace and generosity as you offer support for future generations of women pilots that touches my heart.

Blue skies.

UNDER
THE
RADAR

CHAPTER ONE

Zimbabwe, Africa

The road narrowed at the turn where Tinashe Kagona waited in an unmarked vehicle. He tapped the glowing ashes off the cigarette he held out the window. The smoke rose into the night and disappeared, much as he planned for his target.

Mistakes had been made...dangerous ones he would rectify. Opportunities arose for those with the stomach to take the necessary steps.

His cell phone vibrated. "*Ee yebo,*" he answered.

"Minister Mukono has left the feast. The boy is not with him."

"Impossible. He arrived with his family."

"Yes, sir. No one saw the boy leave, but staff say he was taken home. I'm headed to the house to collect him."

"You understand what is at stake?"

"*Yaa.* I will call when he is secured."

Kagona ended the call and waited for a message saying Finance Minister Mukono had passed the first lookout. Fifteen minutes later that message arrived.

"He's getting close." Kagona nodded at his driver, who

signaled others in a nearby dairy truck. The rough engine on the battered vehicle roared to life. The transmission balked when the driver shifted into gear.

The radio crackled. "Mukono is traveling faster than anticipated. Less than a minute away."

Gears screeched, grinding at Kagona's nerves. "Get that truck across the road." He dropped his cigarette to the dirt. The dairy truck rolled slowly forward down the rough edge of the road. A deep rut along the apron and asphalt slowed the transition. Car lights headed toward the turn. "Damn truck driver. He won't make it in time."

"Sir"—came through on the radio—"the car is coming too fast. Mukono's driver was instructed to take the turn slowly."

"Mukono must be driving. He trusts no one."

Kagona's driver snorted and grabbed a set of handcuffs from the dash. "For good reason, sir."

The foolish dairy driver rushed his efforts to stretch the vehicle across the road, succeeding in overshooting and leaving a gap at the rear.

Mukono took the curve. His headlights illuminated the scene ahead.

Kagona opened his vehicle door, anticipating the quick action necessary. His driver flashed on the headlights. Their beams caught Mukono's passenger—his perfect wife, the very person who had shifted the minister's loyalties. Mukono would give Kagona what he wanted...in exchange for setting her free.

Instead of slowing, the minister gunned his engine, heading for the gap. Fear of failure gripped Kagona. His well-laid plan was about to fall apart.

The car's tires on one side left the asphalt and caught the rut that had slowed the truck. The wheels couldn't escape and the car failed to clear the rear of the dairy truck.

Bumpers crunched and ripped. Glass smashed. The truck shifted. The car spun off the road, finding the uneven ditch to the side. With a disheartening roll, objects flew from the vehicle. Sickening sounds of complications ended in silence with the car upside down.

"Move," Kagona shouted.

His driver sprang from their vehicle and rushed toward Mukono's car. They needed the minister alive. Without his knowledge, all could be exposed. Kagona grabbed a flashlight and followed.

The truck lights backlit Mukono hanging in his seat belt, blood covering his head and face. Fluid dripped in muffled plops somewhere deep inside. A man reached in and checked for a pulse. He gave Kagona a nod.

Relieved, Kagona again focused on possibilities. "Get him down."

A man crawled into the vehicle and lifted Mukono up while another released his seat belt. Mukono dropped into the arms of the men, who dragged him from the car and stretched out his unconscious body.

Kagona's driver, who had field medic experience, checked him over. The minister moaned. "Hospital, sir?"

"How bad is he?"

"He will survive."

"Then proceed as planned. I'll have a doctor summoned." Kagona signaled his men to carry Mukono away.

Another man stood on the far side of the car, looking down. "What about Mrs. Mukono?"

Kagona's driver shifted to her side and knelt. He gave a shake of his head.

Complications. Kagona joined him to assess the situation. The top of the vehicle trapped her lower body against the ground, surely crushing a good part. Cloth and skin covering her arms had been scraped away by the rough ground. A cursory inspection made it obvious without hospital care she would not survive. Even with it, her chances were slim. That was most unfortunate.

He signaled his driver to leave and join those handling Mukono. "Take the license tag with you. The story will come out when I'm ready."

Once alone, Kagona shone his light onto a dark face ravaged in pain. Beauty had been wasted on the likes of her. Pity.

"Help me," whispered past her lips, accompanied by bubbles and a dribble of blood. Large brown eyes with the expectation of rescue stared up at his face, shrouded in darkness behind his light.

"I promise, Mrs. Mukono, you shall have your help." He knelt closer and with slow, purposeful movement swept the light across his face.

Hope fled from the beseeching eyes of the Ndebele troublemaker. She struggled like an impala in a lion's jaw. Blood pooled around her mangled arm.

Without her, it would be more difficult to coerce Mukono's help, but there was little Kagona could do to change that now. He'd use Mukono's son. Rumor had it they both accompanied the minister everywhere. Kagona only prayed the child's odd behavior hadn't alienated his father's affections.

"Is it sad that your last thoughts will be how you were the cause of your husband's demise?" A small hint of pleasure came with his nemesis suffering the loss of something he cherished. "Do not worry, I'll make sure your son shares your husband's fate. He'll be in my hands within the hour."

Oddly, an edge of her mouth quirked up, smiled, perhaps. Did she laugh at him or know something he didn't? His chest tightened and his grip squeezed around the flashlight.

"There is no hope for you, Nomathemba. You will not escape death. The country will mourn that a revered couple has met such a tragic fate."

A strong smell made his next move easy. Gasoline soaked into the dry earth, and as it saturated the ground, a small pool formed and a rivulet flowed toward his feet. He set the flashlight on the ground so the light shone upward, highlighting him reaching into his pocket, removing a cigarette, and placing it between his lips. He patted several pockets before opting for a match pack instead of a lighter.

A grin blossomed as he pulled a match loose. He left the pack open. He held his hands into the light beam and struck the match. The phosphorus sparked to life. The flame danced and swayed, so bright, yet with so short a life. He brought the flame to the remaining matches. They flared to life in a grand rush.

He scooped up the flashlight, then tossed the pack onto the ground near the car. Stark terror shone on Nomathemba's face. A burst of blue flames whooshed across the gasoline-soaked ground toward her.

He backed away.

Even short of breath, she managed a piercing scream. He'd remember it for a long time. Perhaps forever. With a deep breath of dismay at the added difficulties ahead, he pocketed his unlit cigarette and walked away, wondering how long until the shrill noise ceased. A small explosion rocked the night air, making the point moot.

His driver had the engine running, and they fled the scene before gawkers arrived. The phone on the dash rang. He snatched it up. The familiar number indicated the politicians had become impatient.

"Edward Mukono has forced us to change plans," Kagona reported. "I'm afraid there has been a terrible accident." Silence weighted the air about him as the pronouncement sank in.

"Are you saying he is dead?"

"For the moment, he is alive. If he stays that way, he will talk." Kagona described the overturned car and the unfortunate demise of Mrs. Mukono.

"You assured me this would be a simple act. That the Mukono family's disappearance would be easily explained. Much we have risked could be exposed by that man."

"This is simply a short setback. He will give up his information."

"You can't be sure of anything. Did his son survive?"

"He wasn't in the accident. We're securing him now. I suspect in forty-eight hours you'll have the answers you desire."

"There is no job if you fail."

"Likely not for either of us." Kagona ended the call, aware of how her voice grated against his nerves.

CHAPTER TWO

One week later
Hoedspruit Air Force Base, South Africa

Above one of the more unusual air bases in the world, Joni Bell heard an ominous snap from beneath her seat. Something heavy scraped across the floor to the rear of the cockpit. The compact experimental craft's nose pitched up, increasing the sink rate and plunging her toward the earth.

"CG's gone to hell," she relayed to Operations. "Something broke loose under my seat."

Years of flying kept her focused, steady, and on task. Fly the plane. The pitch and nose dropped with necessary inputs to a specialized collective control. The errant object slid forward, shifting the plane's center of gravity again. Joni added power to arrest the sink rate. The craft—Taz, like its devil namesake—wreaked havoc.

So much for a spot landing. She simply wanted to make the runway—alive and not in a shattered pile of debris.

On the other hand, if she went down in a blaze of glory, the dignitaries watching her performance would have something to talk about. Taz might be able to take off like a helicopter, fly as

7

high and fast as a plane, and then swoop in to land on a dime in near silence, but they wouldn't see it now. This compact, one-person future Special Forces toy was about to be splattered at their feet.

The last of the trees and brush from the surrounding African wilds turned to golden savanna grass and then to asphalt as she passed the runway threshold. One of the cheetahs patrolling inside the fenced base tried to pace her for a short spurt, until the chase truck with cameras capturing Taz on video raced down the runway.

She throttled back the push propeller. The overhead rotor whipped around to cushion the landing. With fuel tanks purposefully topped off for a max load, impact came sudden and hard, but miraculously on the gear, bouncing Taz several times. Instead of the expected crunching and tearing of metal, Taz jolted to a halt.

Shaken, Joni forced out a long breath. "Thank you, God."

She unhooked her harness and tore off her headset, taking with it entangled strands of red hair. With adrenaline-enhanced strength, she tilted the canopy attached to the nose forward and leaped from Taz.

The chase truck squealed to a stop. Andre Obergen, the graying South African program manager, bounded out.

"Looks like you had some problems." Mr. Company Politics wasted no time in getting to the point. "The plan was to hit the target."

"I thought about it, but preferred to die of old age." She reached back into the plane before she said something she'd regret and scrounged around on the spartan floor. Far back behind the seat, a black and definitely alien object to Taz caught her eye. "Taz is a prototype, not a production craft. Don't expect perfection yet."

"I don't, but hope your explanation for missing the mark will appease our audience."

Her hand shook in disbelief as her fingers closed around the object. Weights slid off as she yanked the object toward her. She glanced over at Obergen, speculating on its origin.

"Perhaps you should be the one to explain." She hefted up a wide nylon dive belt with a few weights still attached. "I'd say this part is about forty American pounds. Someone hooked it with more weights to my seat." She made no effort to hold back surging ire. She'd risked her life while someone played a stupid and dangerous game.

"I believe they chose the seat to maintain a proper center of gravity. I mentioned at the preflight briefing that extra weight was a last-minute request by our distinguished guest. Maintenance did the best they could." Obergen's face remained stoic when she dropped the weights an inch from his Italian loafers.

"Your shortcut could have killed me."

Obergen said nothing. He'd probably given maintenance two minutes to come up with this shoddy solution. "Somebody had better clear the air about the need for this urgent request or this project will be minus one test pilot." Suspicions unsettled her mind. "Just who *is* this guest?"

"An American, like you, one who has authority to make demands related to Taz."

"What demands and why the rush?"

Tires squealed as a van stopped next to Taz. The entire ground and maintenance crew bailed out. They checked over the craft, pleased nothing was bent or twisted. Rather a surprise to her, too.

The workers backed a tug up to Taz's nose wheel. Obergen caught Joni's elbow and guided her away from the action. "Are you capable of handling Taz with this extra weight in the future?"

She shook free. "Perhaps you should explain what you have in mind with that question."

"We'll cover it at debrief."

"I'm the test pilot. Remember? Taz is a prototype vehicle. Special Forces may want this thing, but it's *not* yet been configured or field-tested for a soldier and his heavy pack."

"Our guest is interested in what this prototype can do, not the next."

"That's ridiculous. This version of Taz can barely carry Kriegler, who's a good sixty pounds heavier than I am, and still have enough fuel for the test flights." The other project pilot was likely appalled at her performance.

"Then think how much farther Taz could go if you flew and we maxed it out with fuel."

"We're not at that stage in the test program yet. What's going on?"

"Best if I let our American guest brief the team. His last-minute arrival didn't provide an opportunity before the flight."

"That's no excuse. The Americans may be contributing to this test project, but South Africa controls it."

Obergen had risked her life for some political bigwig and was bending to his demands. Disgusted, she headed toward a bunker-like hangar the South African air force had leased to the company for developmental testing.

Obergen followed a step behind. "Our VIP was most concerned with Taz being able to land on target. I'll explain how the weights broke loose and yet you landed with no damage. The demonstration has been a relative success."

"If you wanted someone heavier, you could've used Kriegler." She picked up the pace. "He would've jumped at the opportunity."

"You were already on the schedule."

"Death by default?"

"You flew special ops for the US military. I'm sure this was trivial compared to what you encountered."

Sure, but then her crews had been staffed with the sharpest

It's an older picture from his university days, but it's the best we have."

She maneuvered to glimpse the photo. A man with blue eyes and blond hair, refusing to be combed in any particular direction, stared out from the wall projection. From his tanned face and rugged build, rugby was likely his sport, and by all indications huddling in a rugby scrum with him would be highly dangerous.

"His name is Ian Taljaard, pronounced *tall yard*," the speaker emphasized.

Whoever Taljaard was, it was a damn shame these men had his photo. If the speaker's presence had the same impact he'd had on her today, Taljaard was headed for trouble.

Bram coaxed her toward the door, allowing the speaker to come into sight. She froze. Two years of buried emotions returned with a rush. Two years she had spent wiping out the death of a kid because of this man's recklessness and her failure.

She spun a perfect 180.

Kagona leaned over the shoulder of his number-one man, who sat in front of a monitor at Central Intelligence Organization's headquarters. Chaipa brought up a video of a home office, showing a desk and a young boy frozen in the frame.

"Finance Minister Mukono had his home office monitored." Chaipa tapped a key and the soundless video played.

"No trust in his family?"

"His family, yes, but not the home staff. He became lax after they left for the day."

Long shadows from a partially opened window indicated the video had been shot near sunset. A young boy of early school age hovered near Mukono's desk. Several papers lay open, yet the minister wasn't present. The boy grinned and climbed into the leather desk chair. He scooted the chair closer and knelt on it to gain more height. He played with items on the desk before focusing on two pieces of paper.

The boy spread them apart, then leaned over each one for a few minutes before plopping back in his father's chair.

Kagona shook his head. "He looks barely old enough to read."

"He reads perfectly well. Watch. I've edited to show the important moments."

Mukono entered the room and tickled the boy curled up in his chair. The father regained his seat and perched his son on his lap.

The boy pointed to the papers, then spoke to his father. Mukono laughed and turned the papers over. The boy grinned and rattled off something of substance to his father.

Mukono's eyes rounded. He snatched up one of the papers, read it, set it down, and then grabbed up the other. The man grinned with surprise and tickled his son again.

"What are they doing?" Kagona asked.

"They are playing. The boy is repeating content on the papers."

"All of it?"

Chaipa tipped his head with uncertainty. "Parts, from what examiners can tell. His father gives a piece of information. The son finishes off with the rest."

"Anything important on those papers?" Kagona considered the implications of the child's memory.

"Merely a trivial list of names and addresses related to private schools in Harare. But the interesting part follows."

Chaipa brought up another portion of video. On it, Mukono reached inside his cabinet safe on the office wall and removed a document packet.

Kagona had personally viewed the ash remnants from the safe after his men had forced it open. Fools. They'd triggered an incendiary that destroyed the contents. "Could those papers in his hands be the ones destroyed?"

"I believe so."

Excitement rushed through Kagona. Progress at last.

The minister carried the packet to his desk. He raised a long wood bar along the top edge, blocking the desktop contents from being captured by the camera. Minister Mukono knew the possibilities of this video being seen by others. He set to work studying the papers, and in a few places marking something with a pen.

Chaipa jumped the video ahead to Mukono's wife, Nomathemba, bringing her husband tea.

"Pay close attention here." He sped the video ahead until the minister poured and drank the last of the tea from the pot. At one point he sat back in his chair. Then he leaned forward and flipped though more documents.

"Did you see it?" Chaipa backed up the video about thirty seconds and let it play again.

"Stop." Kagona pointed at the screen. "The tea set shifted position, and yet no one touched it. Something has been edited out."

"More than fifteen minutes, but not by us."

"Interesting." Kagona straightened. The implications promised trouble or opportunity. He'd exploit this knowledge to every advantage. "Were you able to retrieve any of it?"

"Not yet. Mukono successfully deleted whatever happened in those minutes. The question is why? It is the only gap discovered."

Kagona toyed over possibilities. One stood out. "Mukono is protecting someone. When was this recorded?"

"After the staff had gone for the day."

"From your choice of video clips, I assume you think Mukono's son saw his papers. It's more important than ever we find that boy. If his memory is that good, we can get all we need from him. But be careful. Our enemies may be after the same thing."

Kagona snatched up a satchel and his phone. "I'm heading to Mutare with our special visitors. When I get back, I'll question Mukono on the video. Find me his son."

"Yes, sir. One more thing. President Tangwerai's office left a message. He wishes an update on Chitima."

"Lie as usual, Chaipa. Tell him all is well. Soon enough we'll have removed the loose ends and be well on the way to lining his pockets."

CHAPTER THREE

Hoedspruit Air Force Base, South Africa

A haze of painful memories pounded in Joni's head as she meandered through hallways and eventually walked up a sloping corridor to a ground-level exit.

Bright light from a small window illuminated Bram blocking the door. "Obergen predicted you might react this way. At least stay and hear the VIP out."

"You've no idea who that man is."

"Actually, I do. Obergen said he's Mr. Paul Lemmon."

"You claimed you hadn't met him." She made no effort to tamp down her anger.

"I said *personally*."

"Well, his name is only half the story."

"Yeah." Bram softened his voice and stepped closer. "Obergen hinted he's American CIA."

"The truth is that VIP makes life hell for anyone involved with him." Deep fury, she'd thought long dissipated, bubbled forth. She attempted to shove Bram aside. Her measly effort had no effect.

"Come on, *bokkie*. The continuation of this entire development program is dependent upon you."

"Is that what Lemmon claimed? He's made coercive threats an art form. Did he tell Obergen that the US would pull their project funding and kill their partnership if the pilots refused to help?"

"You have it wrong. The South African president asked for his help. Come on, hear him out."

"If everyone in that room knew how Lemmon manipulated people, they would scatter for cover. I spent a grueling month fighting to survive in a jungle because of him. I don't plan on giving him another second of what's left of my life."

"I understand your reluctance, and while I might not agree with it, I'll support your decision not to fly. But you must realize these are my countrymen working on Taz. If you choose not to participate, I will. We South Africans are a fierce lot. I'm not going to let them down."

Bram headed back down toward the briefing room.

Damn it. She leaned her forehead against the door. This place had become her new home. They'd accepted an outsider, a foreigner, and made her feel welcome. She'd do the same thing in Bram's shoes.

Regretting the decision, she slowly followed.

Obergen rose when she entered the briefing room and motioned her to a chair. "Glad you could make it. There's a lot to cover and tough choices to be made."

No kidding. The CIA didn't involve itself in anything simple.

Attendance was surprisingly small. Only Mosola Tinibu, the maintenance chief, and Bern Loots, the ops officer, rounded out the crowd. Their faces read of curiosity combined with hope...likely hope for Taz's stable future and their assured paychecks. Heck, the guilt piled on.

With a careful eye toward Lemmon, she dropped to a seat, crossing her arms against the slimy feeling of being around a master manipulator. Oddly, her gaze shifted to the rugged blond man projected on the wall. She hated to think he was associated with Lemmon. What a waste.

Lemmon leaned forward onto the briefing table. "I'll get to the point. Budgets are tight. The American agency funding a large portion of this project has expressed concerns Taz is next on the chopping block."

He gauged their expressions, likely assessing whether anyone had the courage—or rather, foolishness—to question his revelations. "None of us want to end this joint project. The Americans desire to see Taz-type spec-ops vehicles proceed to production. That's why I propose an operational test using the current prototype."

Bram frowned. "I'm as anxious to fly as any test pilot, but Taz isn't in an operational configuration. Its max load tolerance is roughly equivalent to my weight. The new prototype with the expanded payload won't be ready for months."

Lemmon settled back and played with a pencil. "I realize Taz isn't currently configured for heavier loads or night capabilities. However, if Miss Bell flies, we can pack on more fuel."

Joni snorted. They didn't give a damn about talent or skill. "I think what Mr. Lemmon is saying, Bram, is that size matters." Distrust left her chilled. "I should also point out that even with extra fuel, Taz has limited range."

Lemmon shrugged. "We can launch close to the target. It's merely across the Limpopo River."

"In Zimbabwe?"

"The operation is straightforward. Fly Taz into the country. Pick up what we needed yesterday from Taljaard"—Lemmon gestured toward the photo on the wall—"and fly it back."

Zimbabwe was a deteriorating mess before the corrupt, decades-in-power president had died. Now with a new president struggling to grab power, political wolves had gathered to rip apart the system while they battled for control. An upcoming election promised violent controversy.

"I'm no longer flying classified spec-ops missions for the military. If I'm risking my life going into another country without permission, I'd like to know why."

Lemmon calmly studied her, apparently not surprised by her demand. "Fair question. I'd like you to retrieve information related to the Marange region."

"Is this about blood diamonds?"

"Actually, it's about tax revenue related to the Marange diamond fields. Mr. Edward Mukono, the minister of finance in Zimbabwe, disappeared after his wife died in a car accident a week ago. Authorities there claim he is responsible for her death and has fled, supposedly taking five hundred million in tax revenue from the Marange fields with him."

Consider her jaded, but clearly there was more to this story. "Zimbabwe has been ripe with corruption for years. Why are the Americans getting involved now?"

"We believe officials have imprisoned Minister Mukono. Before his disappearance, he passed along information on a few accounts, requesting our assistance to track the provenance. He passed it through Mr. Taljaard, a close family friend. Mukono had further information and directed if anything were to happen to him, Taljaard should get the data to the South African president. Now that Mukono has disappeared, Taljaard requested help to get the remaining information out of the country. With the country's CIO—that's the Central Intelligence Organization—and police searching for it, speed is critical."

His explanation left Joni uneasy. "Taz is for stealthy insertion. Getting a single Special Forces soldier a long way into a region to do business and then out again swiftly and unnoticed. This mission is basic information retrieval. You don't need Taz. You need the internet."

"The majority of the data is still integrated in the original collection system. Even if we could get in to extract it, a few deterrents exist. First, the Zimbabwean government is woefully behind in modernization. Cell phone and internet connections are spotty. Add to that, Zimbabwe's former president contracted the Chinese to build an intelligence complex for their CIO. We believe that complex is capable of signal interception."

"Like Zim's own little National Security Agency?"

"Precisely. Interestingly enough, they parked this new CIO complex right across from the Mazowe Earth Satellite Station."

The irony wasn't lost on Joni. "Mazowe is one of Zimbabwe's international telecom links."

"It connects Zimbabwe's internet traffic to the satellite. I think you can see why we didn't trust transmitting data by phone or internet. They'd likely tap into it."

"If you're trying to convince me not to go, the CIO having advanced hardware adds one more good reason. You also said a few deterrents. What else should I know?"

"The equipment containing the information is delicate and a tad cumbersome. Thus we believe the easiest, quickest, and most secure way is to fly it out." As though Lemmon caught her resolve softening, he added, "The South Africans have become involved not only because Mr. Mukono is a friend with the current president, but because they'd like a stable country on their border."

Bram, who'd quietly listened to Lemmon's arguments, broke his silence. "We frequently use a chopper for chase when Taz flies. Why not make this easy and let me fly it in?"

"That possibility was considered, but set aside. Taz is a much stealthier craft for daytime use across the border, can reach to Taljaard's location, and being so small, it's easily concealed."

"Right." Bram didn't look convinced. "Just shove it in the bushes."

"You could. Taljaard will have a camo net for cover. Taz's presence during the day is much less threatening and unlikely to draw notice. The landing zone was once part of a wildlife conservancy. Even though most of the old landowners and their equipment have left with the turmoil, on occasion odd aircraft like homebuilts or ultralights fly over for animal counts. Ground observers wouldn't think twice about Taz."

Her boss appeared agitated at his pilots' questions. "Think long-term, Joni. Mr. Lemmon is offering us an operational test to prove Taz's field capabilities. Taz is far enough along in the testing phase to risk throwing it out in the real world. If we pull this off, we're guaranteed future funding by both the South Africans and the Americans."

"Let me get this straight." Joni focused on Lemmon. "I would land Taz, Taljaard hands over the equipment—which you are positive will fit into the craft—and then I fly away." Seemed simple enough, and therein lay her doubts. Mr. Lemmon and the CIA didn't do simple.

"More or less. We don't believe anyone realizes Mr. Taljaard has the information, although from his past disagreements with the authorities and friendship with Mukono, his movements are likely being watched."

"What happens if Taz is discovered in Zimbabwe? It could provoke an international incident."

"I doubt it." Lemmon exuded overconfidence. "Finger-pointing and complaining might occur, but the truth is, Zimbabwe needs South Africa, not just for food, goods, and a seaport, but for aviation fuel. If they discover Taz, we'll claim you strayed off course. Your presence will simply be explained as an airspace incursion by a foolish pilot."

Bram actually laughed. "Right. Like a seasoned pilot would stray across the Limpopo River marking the South Africa-Zimbabwe border and not notice."

Joni agreed. "It's more likely one of Zim's Chinese Migs will intercept me in flight and shoot before asking questions."

"I wouldn't worry," Bram added. "Zim pilots crash those more often than fly them."

She looked over at her boss. Since American money had

been infused into the project and they'd hired her to represent US interests, morale had risen, and amazing strides had been made with Taz's technology. That made Obergen all the more vulnerable to meeting Lemmon's demands without thoroughly vetting them.

"You should go into this with your eyes open," she said to him. "If something goes wrong, you'll be left without a prototype and no guarantee of future funding."

Obergen nodded. "Mr. Lemmon and I have had a lengthy discussion on this topic. It's a risk I'm willing to take, if you're willing to fly. Our office is counting on your cooperation." He indicated the maintenance and operations managers, who had faded into the background during the discussions. They both nodded.

Joni rose and paced along the conference table. She ran a hand over her head and turned away, taking a few steps, perhaps to convince herself not to succumb to this reckless plan. Odds were this mission was not about Marange diamonds. Lemmon was holding something back, probably something big.

Positive the great wizard Lemmon was leading the project office down the yellow brick road to Oz, she turned back to the expectant faces, fingers crossed they wouldn't be betrayed when the curtain pulled back on Lemmon's real motives.

"If I crash in Zimbabwe and live through it, I'm walking away from Taz, leaving whatever information is onboard." She looked directly at Lemmon. "Bottom line—when all else fails, I'm saving my own ass."

CHAPTER FOUR

Early the next morning, Joni flew Taz toward the undulating, dirty, blue-green ribbon of the Limpopo River—the physical division between South Africa and Zimbabwe and the last chance to turn back. She avoided Beitbridge, an official border crossing and a hot, dusty city with a reputation for abundant petrol, collecting bridge tolls, and being one of the few places where Zimbabwe kept functioning radar. With no clearances or a filed flight plan, the last thing she cared to do was show up as a blip on someone's screen.

Lemmon had briefed her on Zimbabwean situational politics, the people, the customs, and her contact, Ian Taljaard. Regardless of what Andre Obergen believed, Lemmon and the US—not the South Africans—controlled this mission. Lemmon had waited till near liftoff to privately pass along Taljaard's GPS coordinates—a move she found rather curious.

When maintenance had rolled out Taz, she should have ignored the guilt and been smart enough to walk away. They had changed the registration painted in big, bold letters on the aircraft fuselage. The experimental "ZU" now read "ZS," typical of light aircraft from South Africa, which frequently flew in Zimbabwe.

At least this country didn't have rebels waiting with stingers to bag a foreign spy plane, like they did in Colombia or the Middle East. Even the air defense pilots here would probably mistake her for a South African recreational craft and warn her before blasting her from the sky. At least, she hoped so.

Joni checked the GPS to update her position. Farther into Zimbabwean airspace than Lemmon had originally indicated, she headed northwest, passing over giant baobab trees standing like sentinels on the golden lowveld scrublands. Adrift in a land with a history of violent changes, those trees had survived centuries of wars, politics, lightning, and man. If only she could feel so secure. Lemmon had lied to her before. He most likely had done it again.

Gradually the veld butted against partially abandoned agricultural areas fed by small rivers. Acres of struggling fruit trees and farm furrows buried beneath weedy overgrowth hinted at political turmoil. Lemmon's brief had described the inflation that made food exorbitant. The government blamed it on a past drought and foreign sanctions. The hungry blamed it on government mismanagement and greed.

She shifted in the confining pilot's seat. The country's demise wasn't her concern. She planned to snatch the "paperwork" and flee back across the border.

The lowveld melded into the higher elevation bushveld, where lichen-mottled granite kopjes, weather-polished rock hilltops, and massive outcrops jutted from bushes and trees attempting to engulf them. She drew in a deep breath, inhaling the stark beauty. Zimbabwe had memorable scenery everywhere, not only at the famous Victoria Falls.

Feelings of tranquillity disappeared as a speck appeared in the distant sky. Another aircraft. Her body tensed as the speck remained constant, not moving across her field of view. As seconds ticked away, the aircraft grew to dime size. The lack of movement and increasing size meant only one thing—it was heading straight at her.

She shoved in full power and started a climb while banking on a course perpendicular to the approaching plane. If she were lucky, the pilot hadn't noticed her, and with enough distance Taz would appear similar to other light aircraft. If unlucky, the pilot might well be searching for her. One good look at Taz and a Mig would be at her little wingtips in minutes.

Her skin tingled with desire to turn or dip Taz's short wing to look for the oncoming plane. She fought the urge. Every movement could reflect sunlight and alert the passing pilot. Not knowing the exact type of plane, its true distance away, or its speed, she guesstimated how long it would take a slow aircraft to pass if it were about twenty miles out.

A trickle of sweat rolled into her eye. She wiped away the salty burn with the soft, quick-dry hiking shirt she wore. Obergen had decided a flight suit hinted of military maneuvers and might get her shot by overexcited authorities. He believed a woman forced to land in typical blue cargo shorts and a light pink shirt offered a higher probability of survival. Ha. Who was *he* kidding?

Before reaching an altitude high enough to touch the face of God, she leveled off and did a 180. The other plane was past, but close enough to identify as a twin engine. With only a few miles and minutes until the rendezvous with Taljaard, she backed off on power and rushed the descent.

A tan rock mining scar came into sight on a faraway hillside. It marked the approach to her destination, an old game reserve. Other telltale landmarks appeared as her altitude decreased. As she passed over the burned shell of a brick farmhouse and metal roof of an old barn, a flash of sunlight came from a clearing in some trees ahead. The GPS coordinates were almost dead-on. Mr. Taljaard should be waiting below.

The grass and dirt opening appeared to have ample clearance

for a helicopter and, on a good day with a stiff wind, perhaps for a short-landing aircraft. At the moment, though, any helpful winds remained calm.

Her palms moistened as she powered down the push propeller engine. The big rotor above whipped quietly, powered by its flow through the air. Airflow over the thin wings allowed for a smooth descent. Sleek, lean, and light, Taz now operated almost silently. People on the ground would have to catch sight of the craft to know Joni was landing.

The quiet craft moved easily to her inputs. Her confidence soared. With the fuel low and the remainder evenly distributed in the wings, weight did not hinder the descent.

Taz maintained a perfect glide slope as they rushed closer to the clearing. With one bounce, she settled onto the reserve.

"Perfect." Joni unhooked her harness and snatched off her headset. Now to grab the info, add enough fuel to get back across the border, and then fly the hell out of here. She straightened her ball cap that read "Vic Falls."

After the rotor stopped, two men appeared, both with dark complexions, thus not fitting Taljaard's description. They rolled a fuel barrel from the underbrush toward Taz.

Something moved in her peripheral vision as she tipped open the canopy. A man stepped out of the bushes. There was no mistaking the face on the wall—well, almost. Gone was the youthful glow of college days, replaced by the rugged, determined set of a man confident of his role in the world. Dressed in traditional southern African khaki, Ian Taljaard appeared taller and a bit more imposing in person. Strong thighs protruded from shorts and melded into long, athletic legs, ending in practical suede veldschoen field shoes with thick black waffle tread. His blond hair had lengthened since the photo and stuck out from under a wide-brimmed safari hat.

Joni stepped out of the cockpit, evaluating him through aviator sunglasses.

He dropped the day pack slung over his shoulder. Those intense blue eyes from the photo narrowed, and his gaze tracked across Taz and then her, looking inexplicably irritated and not at all pleased.

"Bloody hell, they sent a woman." His deep voice resonated with a blended British-European accent.

"Breasts and all."

"I noticed." He didn't grin. "You're late."

"I had to avoid traffic not far from here."

"What type of plane?"

"Twin engine. I wasn't close enough to see color or registration."

Taljaard visibly relaxed. "It's likely Gundermann. He runs safaris out of South Africa for wealthy clients. Not many other locals up these days. Fuel is hard to come by."

"I'm not sure if he saw my aircraft or not."

"Half of the hunting Gundermann does is illegal. He'd be the last one to report you. He's probably concerned you saw him." Taljaard pointed at Taz. "Close it up and follow me."

He snatched up his pack and blended back into the brush as smoothly as an animal familiar with this land. His rapid departure surprised her, but not his movements. Born in the mideighties, he had grown up in this area, gone to Oxford for university, and then returned to Zimbabwe to work. From his bio, she knew few of his childhood classmates came back after school. Opportunities here for whites had dwindled after the country gained independence in 1980, and had become downright dangerous for many farmers and businessmen over the last fifteen years. Taljaard must have had a strong reason for returning.

"Eh, miss?" A man with the fuel barrel stood next to Taz, not sure where to refuel and looking at her for direction.

"It's near the fuselage above the right wing." She stuffed a satellite phone in a deep pocket on her cargo shorts, then lifted the canopy into place.

She found Taljaard under a leafy mopane tree, his arms folded in obvious impatience and a veldschoen tapping the reddish-brown dirt under his foot. "I'm assuming Lemmon sent someone capable of fieldwork," he said. "What can I call you?"

A laugh almost bubbled from her lips. Did he expect her to use some covert name? How about Jewel, Susie Q, or Lara Croft? "It's Joni."

He nodded. "Time is critical. We've a way to go before dark. I hope you're armed."

"I damn well know time is critical, Mr. Taljaard. I plan to be back in South Africa by dark. I didn't come to follow you somewhere or take part in any covert action. I came for the information in whatever configuration it's in. Hand it over and I'll get the hell out of your life."

"It?" His eyes rounded as though incredulous at her words.

"Yes, the paperwork, electronics, computer drives, whatever you want rescued. All sixty, eighty, heck, a hundred pounds of it."

Taljaard stared hard at her as though befuddled by the words or at least at a loss for how to respond.

A moment of clarity struck. "Lemmon didn't give me the whole story, did he?" She dropped her head into her hands, too tired of CIA games to let fury take over. With a disbelieving sigh, she looked up at Taljaard. "I'm outta here."

Taljaard let his frown roll into a desperate smile as he grabbed her arm. "It was too dangerous to bring the information with me. Some difficulties have arisen. I had to confirm your arrival before retrieving it."

Feeling ill at ease with the entrapping touch, she shook loose from the heat of his grip. "How far away is it? In the barn I saw? Or a mile down the road?"

At the shake of his head, she added. "Ten miles? Twenty?" All indicators waxed iffy for a quick departure. If Lemmon had realized this possibility, then he took a risk exposing Taz.

From his knapsack, Taljaard snatched out a neoprene water bottle for a drink. "I'd say about an hour without any distractions." He offered her a drink, but she waved him off.

"And how high is the possibility of encountering these *distractions*?"

"High." He took a quick gulp of water. "But manageable if you stay with me." Taljaard spoke with such matter-of-fact confidence that she nearly choked. This was a wildlife reserve, not a jungle. Tarzan didn't need to swing her though the canopy.

"I'm not going with you." She glanced at her watch. "I'll give you an hour to get back before I leave."

"Won't work. There are too many uncertainties. I'll have you back in time for an early-morn takeoff."

"Morning? I can't leave Taz here overnight. Every hour I stay, the risk of discovery increases."

"Taz?"

"My aircraft. The implications of its discovery are frightening."

"No one knows it stopped here. As expected, the landing was almost silent."

She bit her tongue before asking how the hell he knew enough to *expect* the landing to be quiet.

As though sensing her discomfort, he added, "My two mates are adept at camouflage. No one will discover Taz." Taljaard acted like he spoke simple truths with which no one had yet disagreed.

She tucked in the loose tail of her buttoned shirt and looked back through the shrubby trees toward the clearing. "I'd rather wait. I'll sleep in the plane if I have to."

"Oh, you will." Sarcasm dripped from his words. "The big five roam this territory."

"Big five?"

"Elephant, rhino, leopard, lion—"

30

"I get the picture. How about your friends? Aren't they staying to guard the plane?"

"People don't prowl these parts in the dark. My men will return in the morning. You did bring a sidearm?"

Hell, no. She might be a proficient marksman, but an armed woman in an experimental aircraft from South Africa booking around Zimbabwe only gave authorities one more reason to bury her deep inside a dark prison if caught. "We were unsure the weight of the material you wanted rescued. So every ounce is accounted for on Taz."

"In other words, you're unarmed. Shame. Lemmon must have had a damn good reason for sending you." The way Taljaard eyed her said he had yet to discover that reason. "How much weight can Taz take?"

A half-ton gorilla if she wanted a spectacular spinout to a crash landing, assuming, of course, Taz ever got off the ground. Maintenance had stripped what few extraneous parts they could from Taz for extra fuel weight for the flight and to provide a greater margin of safety. So, to calculate an acceptable risk…take away the extra weight she'd crashed landed with yesterday, add a margin for error, and convert from pounds to kilos. "Thirty-five to forty kilos, depending on how much return fuel I take on."

"That'll do. I'll be back by morning. As long as you've a light, you'll be fine."

Her fingers drummed against her crossed arms. He knew damn well she didn't have a light and after his *lions, and rhinos, and leopards, oh my* story, she had no plans to stay alone on the reserve.

Finished with the games he seemed to be enjoying, she asked, "Where exactly is this information?"

"On another nearby ranch. Until the last ten years, twelve ranches connected together as a single safe haven for wildlife."

"And the place we're going?"

"Used to be one of the twelve."

"Used to be? Isn't that a little dangerous?"

"The world's a dangerous place."

He strode off. Like it or not, she had little choice but to follow.

CHAPTER FIVE

Mutare, Zimbabwe

The misty Bvumba Mountains, with granite knolls covered in heavy vegetation, offered relaxation and serenity, but they left Kagona's mind racing and feet itching to return to CIO headquarters. Gravel crunched under the small tires of his golf cart as he headed toward the eighteenth tee. His visitors, playing on the Leopold Rock Championship Golf Course, hadn't answered his calls.

He drove onto lush grass leading up to the tee. He'd no time for appeasing these men with leisurely games, not with Minister Mukono's interrogation going poorly and President Tangwerai's constant need for status reports. Yet his presence with these two foreign businessmen was necessary as their two countries moved into the delivery stage of the secretive Project Chitima.

The last few days, the men had been insistent on inspecting the product and all avenues of transportation associated with it. They didn't trust Kagona's country any more than he trusted theirs. He found solace in that fact. It made them the perfect trading partners, and the West hated them both.

Mr. Rajaei and his assistant prepared to tee off. They

laughed, babbled in their foreign tongue, and one nearly fell over teeing up his ball. Intoxicated. Being away from home and religious oversight allowed vices to take hold. A good time for Kagona to pry for information.

He checked his phone, expecting an important text. Nothing. He shoved it back into a pocket in time to hear the whack of a driver against a ball. Rajaei's ball hooked to the left and rolled off the fairway into tall grass. His partner stayed midfairway, but the ball didn't go far.

More precious time promised to disappear.

Rajaei joked with his assistant as they walked toward their cart. He dropped his driver into a golf bag on the back end and caught Kagona's scrutiny.

"No reason to look dour, Mr. Kagona. This day has been productive."

"I'll save my celebration for after the product has been shipped. Time is short if we plan to meet deadlines. Any delays and we will be visible to our enemies."

"Do not be so concerned." Rajaei took a seat in Kagona's cart. "Chitima will pave your future in gold." He pointed toward the edge of the fairway where his ball had disappeared.

In silence, Kagona drove toward it. Rajaei skillfully handled the political aspects of Project Chitima, and his assistant the scientific, although their true occupations likely lay in the intelligence world. The men had spent several days in northern Zimbabwe at the mine inspecting the product quality and mining operations. Yesterday they had shifted their attention to the eastern border of Zimbabwe and the train yard in Mutare.

While Rajaei searched for his ball, Kagona tapped his fingers on the steering wheel.

"You believe we squander your time." Rajaei found his ball tucked deep against a weed. "An incorrect assumption. My assistant and I have had privacy to discuss the results of our inspections."

"I hope you've found everything as promised." Zimbabwe held the resource Rajaei desired...a resource few other countries were willing to sell to his.

"We are pleased at the quality. My assistant also examined the shipping documents with a close eye. Our concern is the transport through Mozambique. The stretch from Mutare to the port at Beira could cause problems. Have bribes gone to appropriate officials?"

"I have a contact handling that portion. His services have been efficient in the past. I've been assured the necessary inspections will go smoothly."

Rajaei propped up his ball. That the man cheated came as no surprise. "I suggest the initial shipment be personally supervised. There is no reason for *little* things to interfere with our deal."

"Be assured, all is as requested." Kagona tired of feigned politeness. "I'm not one to drop my guard. If light of this deal becomes known, a number of countries would be rather hostile toward us. Certain forces might attempt anything to make this endeavor fail."

Rajaei's hit arced across the fairway to long grass on the other side. "Be patient, Mr. Kagona. I will sign the final approval tonight. Once the first shipment is at sea and headed into our hands, foreign powers will see there is little they can do. Excuses are our specialty."

The assurances did little to ease Kagona's mind. If Rajaei learned what the CIO had discovered on Mukono's security tapes, he'd be less optimistic.

A deep burr of chopper blades announced their transportation had arrived.

The visitors did a poor job of completing the hole. While they freshened up at the clubhouse, Kagona checked a text message delayed by Zimbabwe's ineffectual phone system.

The phone sat weighty in his hand as he read the message twice in disbelief.

Asset on way to make collection. GPS destination unknown. Taljaard key. Find him. Find your resource.

Taljaard. Kagona's heart raced. Bastard. He'd been problematic off and on for years. Lately, though, he'd gone quiet. Obviously, too quiet.

The news added complications at a critical time. A week ago, Kagona had interviewed Taljaard, hoping he knew the whereabouts of Mukono's missing son. Taljaard claimed ignorance. Because of their past difficulties, Kagona's operatives had tailed Taljaard south to Bulawayo before losing him.

"Problems?" Rajaei's assistant sauntered up.

Kagona pocketed his phone. Taljaard left a bad taste in his mouth. "A message from the flight crew. They're ready for us."

The assistant's gaze traveled to Kagona's pocketed phone, then back to his face. The man showed no expression, but it was clear he didn't believe Kagona. Any postponement on this deal by Rajaei could cripple Kagona's candidate's bid to unseat the current president.

"I'll have the vehicle brought around." Kagona walked away, needing privacy to make a call.

He rang the right person to help and related the news. "Put all assets into finding Taljaard. He's likely in Bulawayo. I'm returning to Harare first before heading there. Alert authorities to watch for him and any aircraft in the area. Expand roadblocks near the game reserve where he grew up. If seen, do not approach. Contact me and wait for my arrival."

Joni stayed on Taljaard's heels as he cut through brush and grass-covered ranchland. Not sure what wildlife hid in the surroundings, she heightened her senses to the fullest.

Eventually Taljaard pointed to a dirty white Isuzu truck parked near the deteriorating barn she had seen from the air.

"Transportation," he announced with his first words in the last ten minutes.

Hope blossomed. If this trip went quickly, she might still manage a return to Taz before dark. She wasted no time climbing in. Its dust-coated paint blended in like the country's wildlife, which she'd seen no traces of on their short hike.

She did her best to recall Obergen's briefing about the area. Ian's father had run a conservancy ranch. His son had worked for him as a ranger, wildlife manager, and safari guide. A little conversation about his favorite topic might thaw their strained introduction.

"How did this conservancy get started?"

Suspicion narrowed his eyes before he softened to her effort. "Poaching in the national parks nearly wiped out endangered populations back in the seventies. So the government convinced twelve farmers and ranchers to join together and create one hundred and sixty thousand hectares of conservancy. They knocked down small internal fences and erected rhino fencing around the entire thing. Took years, but worked."

"Is the conservancy still operational?" They bounced over the uneven road. A rifle behind her head jostled in the gun rack. "I take it from the condition of the farm back there, things haven't gone well."

"It's complicated. The conservancy was a success." Defensive tones rang in his voice as red flared from his ears down his neck. "The black rhino population nearly doubled. Enough game roamed the property that hunting and photographic safaris provided a stable income for this area. We kept the animal population healthy and in check."

"And then politics got in the way?"

"Land reform." Taljaard swerved to miss a pothole, tossing her against him—a convenient way to end a discussion.

"Bet you try that with all the girls." She attempted to extract herself without touching him further. His face remained somber.

She scrounged for a seat belt, finding only a half. "So what changed?"

"When the government saw their power weakening and a strong opposition party rising, they sped up the process of land reform—returning land to the population living here before colonialism. Bureaucratic greed got in the way. Leaders, their friends and tribe, most unskilled at farming and ranching, received the biggest portions of land. It destroyed the agricultural and wildlife part of the economy. Recently the government has gone after businesses using a similar method. Instead of a controlled process, everything in this country is smash and grab. Not a pretty sight."

"Are we headed to your ranch?"

"No." Taljaard stared out the front window, his jaw locked tight and eyes narrowed.

"Confiscated?"

"*Yebo.*"

His answer likely meant yes in one of the tribal dialects. Although a good portion of the population spoke the official language, English, they also spoke Shona or Ndebele or local dialects depending on their tribal affiliations. "It would seem a bit shortsighted to risk the animals. Aren't they the main draw of tourists and money to the country?"

"Foresight is not governmental policy these days. I'd say seventy percent of the wildlife roaming our original conservancy is gone. Several new owners do nothing more than use the old reserve lands to raise cattle." A smile curled his lips. "A few big cats have figured out how to avoid farmers and feast on the captive prey."

"What about all the lions, leopards, and rhinos roaming the night?"

"Few and far between. You have to know where to look for them."

"I would've been fine back at the plane."

"It's not animals that worry me. It's humans. There are dangerous groups and youth brigades that roam these parts. Harass the locals. The groups become more active closer to elections, particularly because this part of the country is not supportive of the current regime."

She remembered this threat from her briefing. "Are they one of those difficulties you are concerned about?"

"They're worse than stampeding elephants, crushing everything without stopping to ask questions. Many have died at their hands."

"Don't the police do anything about them?"

"Many of the local forces have been replaced by those loyal to the ruling tribe. Others are afraid to act against them and lose their jobs. I was surprised Lemmon sent a woman. Beatings and rape are part of their staple tactics used against their opponents."

"Delightful." Subconsciously, she wrapped her arms tight against her body. "So are any of the original twelve ranches still functioning as part of the conservancy?"

"Four."

"Why aren't we going to one of those four? Wouldn't they be a safer place to hide the information?"

"I can't risk giving the government any reason to destroy what's left of the conservancy."

Even in troubled times, Taljaard had principles and admittedly honorable traits, but ones likely to get her into trouble. Men like him put *their* priorities first. She wanted the info and to head home. He, on the other hand, had the info, but also some necessity to bring her out to a ranch and no intention of revealing why…yet.

She understood at last why Lemmon insisted she carry a fake identity card, one showing her as Joni Taljaard. Lemmon knew the stakes and the risks. Obviously they were high.

"So where does your family live?"

"My aunt and her husband farm in Zambia. My mother lives on a wildlife reserve with my uncle in South Africa."

She started to ask about his father, when Taljaard reached behind her and dropped the rifle from the rack down behind the seat. He then shoved his day pack under the seat, using his foot.

He tilted his head toward the front window. Bad vibes trickled up her spine as four men, armed with rifles and metal pipes, stepped out from under a large tree and into the center of the road.

Alert but calm, Ian stood with the pilot in front of the Isuzu. While initially surprised a woman had been sent, he understood the Americans' logic. She was capable, yet nonthreatening, and considering the information to transfer, a psychological plus. He trusted specialized training had sharpened her observation skills enough she'd recognize the moment to act against these men, if necessary.

Joni tucked her sunglasses into a pocket and removed her ball cap. She wiped a forearm across her forehead, and then used the cap to shield her mouth. "Are you sure stopping was a good idea?" she whispered, before sliding her hat back on.

"Best to stay under the radar for as long as possible."

Two of the four men wandered around the truck, stopping to shift through rubbish in the bed. They lifted out an empty jerrican and two folded feed sacks. Disappointment was apparent as they relayed the news to their companions in a tribal dialect Ian didn't recognize. Not local. That would play in his favor.

His patience had been taxed acquiescing to the men's demand to get out of the vehicle and then answer endless questions. Those questions had revealed their encountering this roadblock had been his piss-poor luck and not part of an all-out manhunt. These thugs were here to remind the locals in opposition territory which political party ran the country.

Unable to hide his obvious fitness compared to their scrawny builds, he backed away a few steps, giving the two men guarding him room to relax behind their weapons. A mistake he planned to exploit.

Joni stayed in place with arms crossed. She studied the men's movements, in particular the one leading the pack who walked back toward them.

"You have no business on this road." The leader repeated his tired mantra. The man wore a dirty red shirt and held his smirking face high.

"Mr. Motobo has the ranch down the way," Ian answered for the fourth time. "He has asked for my help with an injured cow."

The leader frowned, wanting something for his efforts. He tossed his rifle to a companion. With a smirk, he pulled out the identity cards Ian had handed over when first stopped. The man slapped Joni's card repeatedly against his palm. She tensed, likely afraid he'd recognize it as fake.

The man stepped closer to her; so did Ian. The nearest man with a rifle pointed it at him, halting Ian in place.

A grin broadened the leader's lips. "What business does wife have with Motobo?"

"We're partners," Ian answered.

The leader took another step closer. Joni backed up the remaining few inches until stopped by the truck bonnet. One of the other gang members spoke in his native tongue. He sounded irritated with the leader and perhaps pleading. Evidently unsure of Ian, he wanted to head off trouble before it started. At least one in the group had some sense.

The leader snapped back at his companion, reiterating his position of authority. He refocused on Joni. "Why you not flee to Zambia like other farmers and their women?"

Joni tilted her head as though startled by the personal question. "I'm Zimbabwean." She straightened and stared into his face. "This is *my* country, too." By God, she sounded as strong and solid as a real third-generation rancher.

The man reached out and removed her hat. Joni's eyes widened at the advance into her personal space. "You don't look like farmer's wife."

The leader spoke to her in his native tongue. No one needed a translation to understand his intentions. He tossed her hat at Ian's feet as though daring him to intervene.

Be careful what you ask for, bastard.

The man stroked the red hair hanging to Joni's shoulders, watching her face for reactions. Something Ian couldn't read flickered in her eyes, but she stayed focused.

The leader laughed and grabbed toward Joni's shirt. His hand never landed. She shoved against the truck and plowed into the man, leading with an elbow toward his jaw. He deflected her strike, grabbed her arm, and they tumbled to the dusty ground.

Ian's positioning worked perfectly. The distracted man nearest him with the rifle became the perfect weapon to fling into his other two friends. With simple leverage, Ian rotated the rifle into his hands and knocked out one man before the others realized what had happened. Another had dropped his rifle in the jumble and struggled to pick it up. Taljaard swung his foot in the direction of the third man, who fled, leaving the weapon behind. The remaining thug, looking down Ian's rifle barrel, threw down his threatening pipe and overtook his friend.

The leader realized too late who posed the real threat. He tried unsuccessfully to hold onto Joni while he staggered to his feet. He faced Ian. A nervous grin arose on his face along with a knife he slipped from behind his back.

Ian didn't give him time to contemplate using the weapon. The man lay unconscious in seconds.

Joni had the truck running before Ian turned around. He grabbed her hat and tossed the rifles, knife, and pipes in the back before he bounded in the passenger's side. She gunned the accelerator.

"You didn't stay to help me out back there," he said, hardly winded and a bit energized by the experience.

"Help? What do you call this?"

"A wise move."

Her eyes narrowed and pale cheeks glowed red from the heat. "Where did you learn to handle men like that?"

"Dealing with poachers." He considered plopping her cap on her head, but instead handed it over. "They're desperate men."

"I've seen techniques like those before. You could have easily killed them all. Yet, you placed blows meant to incapacitate, and every strike was effective."

"You only get one chance. I've learned not to miss."

"I think Lemmon left something out of your bio."

While her movements showed measured skill and confidence, her questions left him a bit confused at the brevity of her briefing on this assignment. "At least you had a bio to read. Your people didn't reciprocate."

"Look, if Lemmon told you he was sending an operative, he lied. I'm merely a pilot."

"All that matters is you think on your feet and had no trouble understanding you were to make the first move."

"What I understood was you wanted to get out of there without a scene. That nasty guy changed the dynamics."

"And so did you, quite adeptly." No matter what she claimed, her handler, Mr. Lemmon, had assured him he'd send a qualified operative adept at surviving and thriving in dangerous conditions. "Nice try using a British accent back there, even if you came off sounding like an Aussie." At last he got a smile, a wry one, but he deemed it progress. "Those men weren't linguistic experts, but they'd have recognized an American voice and known you weren't local."

She tugged the cap on with one hand. "I could get used to working with you, if I had a death wish."

A motorbike rumbled toward them on the road ahead. "Stay to the left," he commanded.

"I may be American, but I realize what a steering wheel on the right side means."

"And what's that?"

"That the British successfully screwed up driving in most of southern Africa. What if those men back there turn over our identity cards?"

"They won't." He handed over her card. "Collected it off their unconscious leader."

"Will they report what happened to the police?"

"Perhaps, after they lick their wounds and figure out how to turn their failure into some kind of success story. That will take time. They're not from this area so want to maintain a tough presence."

"I hate political unrest. I've met their counterparts in Colombia and hoped to never run into their type again. Do you think they'll remember our names or where we're headed?"

"I made up the rancher's name. The cards are meaningless. The leader couldn't read. I could tell by the way he checked our photos. However, my description will suffice if they talk to the right people."

"And who are these right people?"

"Those who might have figured out I have the information they seek."

"All the more reason for me to get out of town sooner. Get me that paperwork and I'm gone."

Ian settled back and smiled. "You're different than most."

"Most what?"

He ignored her question and pointed down a sliver of dirt

road between two tilting gateposts. "That way." He checked back down the road they had left. Tension in the air hung silent and heavy.

She jerked the wheel and barreled onward. "How much time do you think we really have?"

"I've lived in this area all my life. The locals know I'm not married. Prepare for trouble."

"What kind of trouble?"

He studied her for a moment assessing how much to reveal. "I need to get you a weapon."

"I'm starting to believe the lions and leopards were a safer option."

"They were."

CHAPTER SIX

Joni hadn't driven far on the narrow dirt road when the broken
remnants of a one-lane bridge exposed a car-stopping gap.
The bank on either side of the bridge sloped steeply to a trickling
creek and showed signs of past traffic. "I assume there's a place
to cross, but I'm not exactly adept at fording African creeks.
You're welcome to take the wheel."

"No need. From now on, we go on foot." Taljaard pointed
toward scraggly undergrowth beneath two large trees. "Park in
the brush over there."

She slid the truck between two large bushes, one pressing its
stubby branches against her door. Taljaard squeezed out his side
with the rifle and she followed, joining him as he threw leaf-
covered branches, vines, and dried brush on and against the
Isuzu for camouflage.

"Good enough," he announced as she tossed long grass
against a back wheel.

"I can still see a few spots."

"No more time. It's not safe to stay here. Head to the
creek."

Remembering the guys back on the road, she didn't argue.
But, as she scrambled down the dirt bank, she checked back on

Taljaard. Still at the truck, he brushed away the grass and exposed the tire.

She tamped down rising distrust. A hundred feet up the creek she discovered a fordable point for vehicles. Past jobs for Lemmon had taught her everyone had their own agenda. Taljaard better have a damn good one.

He hustled down the bank and onto the rounded stones covering the nearly dry creek bed. "Stay on the rocks. I'll follow you."

She obeyed, aware he chose the rear position to keep an eye on her. Over the next half mile, she caught him stepping in the soft sand between rocks. He wanted to be found.

The creek bed eventually narrowed, slowing their progress. They worked around driftwood lodged between larger boulders washed in when the creek ran fast and deep. A massive tree undermined by erosion lay with its roots vertical and trunk jutting across the water. With difficulty, she pulled her body up and over, scraping her stomach and limbs in the process.

Taljaard stopped on top of the log. "Give me your boot." He pointed at her feet and held out his hand.

"You're kidding."

"I'm not. Hand it over."

Several scenarios, none favorable, ran through her mind. Lemmon had better know what in the hell he was doing, teaming her up with Taljaard. "I'd like to know what you want it for."

"Trust me."

"Why should I? Nothing is going as planned."

"Everything so far *is* as I planned. Perhaps you chose to trust the wrong people."

Ah, hell, if she had to pick whether to trust a complete stranger with unknown motives or Lemmon, she'd pick the stranger every time. With wary trepidation, she untied and tugged off her lightweight boot.

"Thanks." Taljaard exchanged his rifle for her shoe, easing

her concerns. He hustled across the log and leaped to the far bank, landing on a flat rock. He turned back and pressed the toe of her boot into the sand short of the rock. With surprising agility for his size, he jumped back to the log and after four bounding steps stood above her.

"Why couldn't you simply have told me what you planned to do?"

"Because there might be a situation where I need you to act and not question. Now, we head back," he said, dropping her boot into her hands.

"Back? I just climbed over this log." Not knowing where they were going or even a hint of which direction to head gnawed at her patience.

He bent and offered a strong hand. She ignored it and tugged on her boot, lacing it tight before taking up his offer. He displayed little exertion as he hoisted her to his side and placed a steading hand at her waist.

"How far?" she asked, handing back the rifle.

"To that tree about fifty meters behind me. The big one with branches spread out over the edge of the creek."

She started to understand his plan—lead their followers in one direction, but actually disappear in another. Not exactly novel in the deception game, but depending on those following, it frequently worked.

"This time, step where I step." Taljaard slid off the trunk. "A chimp could follow your trail."

She bit her tongue before it rattled off a petty defense. She thought she'd done a decent job of leaving no tracks. "Apparently, you've done this before."

"A friend and I used to challenge our mates to hide-and-seek on each other's ranches. We got adept at losing them." He stopped beneath the tree and slid his knapsack from his back to his arm. In seconds, he dug out a nylon rope with big knots on one end and a weight on another. He pitched it over a sturdy branch. "You first. Try to drop off as close to the trunk as possible. Keep hold of the rope."

She scooped up the knotted end and backed into the creek. Water pooled around her soles. With quick leaps from rock to rock, she launched into the air. Her boots caught above a knot, which provided support. Taljaard pulled her higher as she swung up over the bank and dropped near the trunk. The nylon rope burned her hand as she struggled to maintain a grip. Without having to be told, she tied her end to the trunk and sank her weight into the rope as Taljaard swung to her side.

"You catch on quick." He untied the knots, and she freed the rope from the trunk.

"I've played on a few tree swings in my time. Did you teach this to all your friends?"

"Nope. Just my best mate." With a snap of Taljaard's wrist, the rope slithered off the branch to the ground. "She was even smaller than you."

"She?" Joni rather imagined him wrestling around with guy friends.

"Yep. She lived near our ranch when I was growing up." Ian wrapped the rope the length of his forearm and around his elbow.

"A girlfriend or more like a sister?"

"Can't say. Never had a sister. But we were close, like mates are." He stuffed the rope into his pack and started off through the brush. Oddly, his last words had crackled with emotion.

She didn't plan on letting an opportunity to delve beneath his elephant hide escape. "What's her name?"

"Nomathemba."

"Sounds—"

"Tribal. Ndebele. Her name means hope."

Had Nomathemba played a romantic role in Taljaard's life? The fact he cared deeply about someone pushed him a notch above the typical operatives Lemmon used. They usually rated money or some political agenda higher than people's lives.

"How did you get to know her so well?"

"Her father managed our reserve and one other." He nudged her toward a rock outcropping of lichen-covered granite. Sudden movement on the rocks stopped her in place. A creature the size of a guinea pig chased another before routing a third from its leisurely basking and sending it darting off.

"Dassies," he said. "They get used to people. Just don't appear to threaten them." He stepped closer and waved an arm. A dassie, camouflaged against the granite, let off a shrill scream. A half dozen of the creatures dashed into crevices.

Curious, she scooted closer.

"No time for observation." Taljaard tossed her a water bottle. "We've a good way to go."

Surprisingly thirsty, she downed a third of the bottle and licked the remaining drops off her lips. Taljaard smiled at her action before looking away. Damnation, if he wasn't enjoying her struggle with the elements. Irritated, she handed back the bottle.

He took two sips and packed it away. "Ready?" He headed off across a grassy field without waiting for an answer.

Joni found herself winded at the pace he set. "Where is Nomathemba now?"

"She married and moved to Harare."

Taljaard signaled the subject closed when he picked up the pace, leaving her behind. After a good five minutes, he stopped to let her catch up and rest. He cast a quick eye from Joni's head to her toes, hesitating at the collar folded open on her sweat-covered chest. "You look hot. Am I setting too fast a pace?"

"I'm doing fine." Her sunglasses hid the lie. The heat ate away her energy. "Let's keep moving."

He stretched his shoulders and adjusted his pack. "It's a ways yet."

"I'm game, but if you need a rest…"

He cocked her a sideways glance and stepped out. She set

out after him, plunking one foot after another across the thigh-high grasses of the field.

"Is this land part of the conservancy?" she asked.

"The conservancy is fenced and starts past that baobab tree in the distance. There's a gate to get inside, but we'll go the quicker way along the outside."

The baobab tree gave her a destination, and achieving it would put her one step closer to the information. She picked up her pace. The true girth of the ancient tree grew evident as they closed the distance. Wide enough to hide a Volkswagen bug, its eerie branches jutted to the sky like roots of a tree. When they arrived at the lone sentinel's side, she noticed two large white blossoms starting to open among new green growth that softened the tangled limbs.

Behind the tree, fencing made of strange kinked wire stood several feet taller than her head. She had trouble imagining it would stop elephants and rhinos, but perhaps new materials made it stronger. She reached out to inspect the strange kinks.

Taljaard shot out his hand and locked his fingers around her wrist. His speed made her wonder about trained reflexes. "It wouldn't kill you." He let go. "But the shock's unpleasant."

"It's electrified out here?"

"Solar power. Low voltage. Enough to keep poachers and locals from dismantling the fence, and to discourage the animals from crossing over."

"Why the funny kinks in the wire?"

"To absorb animal impact." He offered an encouraging grin. "Only a mile to go. You holding up okay?"

"I could use some more water."

Taljaard dug out a fresh bottle. "Make it last."

Before he could toss it to her, he cocked his head up to listen. "Ruddy hell."

In a move somewhat akin to football or perhaps rugby, he bent and grabbed her around the waist, moving her twenty feet backward in a mere second, practically slamming her against the baobab tree. He covered her body with his.

"Sorry, but a little camouflage is necessary. Your denim and pink don't blend in well in this environment."

Between gasps for air, she heard an engine drone overhead. Even covered with new leaves, the rootlike branches of this baobab offered poor cover from an aircraft. "Tell that to the powers that be. They decided I should look like a tourist."

Ian lifted his head, listening. "Can you tell where it's coming from?"

"South."

Still holding her close, he scooted them around the fifteen-foot trunk to the opposite side.

She tried not to breathe as every inhale pressed her breasts tighter against his chest, making her very aware of his close presence. He seemed unfazed, yet the move of his hand from her bottom to her waist perhaps indicated he, too, appreciated their intimate position.

The plane passed the baobab tree low and to the west, so they inched around to the east. The drone softened and eventually the sound disappeared. Taljaard released her and picked up his hat and water bottle that had dropped during his hasty tackle. Her body seemed a bit disappointed at the freedom.

"Recognize the plane?" she asked, brushing tree debris from her shirt and shorts. The single-engine prop had flown low, too low for any casual pilot transiting the area.

"Doesn't belong to any of the typical guides or hunters who work these parts. I suspect it's not friendly, so I'm afraid we've a change to our plans. It'll add another half mile to the trek, but there's more cover inside the conservancy." He took a few swigs from the bottle she'd earlier chugged, and then handed it over. "Drink up. You'll need it."

Thirst sucked water down her throat. She stopped before drinking it dry. They'd need more later.

Taljaard swung his pack over a shoulder and adjusted the rifle. At the fence gate, he punched in numbers on a cipher lock and led her inside.

"The grass is thick and consistent here," he said. "Our pathway might show from the air. We'll follow the fence perimeter until we reach a more suitable area for travel."

She walked with an ear attuned to the air and eyes focused on a tree line across the savanna, where she assumed they were headed. This region stood at the merger of granite outcrops and knolls, trees and heavy bush cover, and dry savannas. The beautiful landscape made her long to forget the country's unstable government.

"Isn't it about time you tell me what's going on? Why is this information so hot?"

"In America I believe you call it *need to know*." Taljaard appeared to know more than a typical wildlife manager should.

"At this point, I'd say I'm in need."

"Maybe, but not at this time."

"What in the hell do you mean, not at this time? Who are you to put stipulations on people here to help you?"

"I'm the guardian of someone else's life. That person is dependent upon my discretion."

Each dribble of information she gained developed a picture totally foreign to her expectations. The more complications Taljaard added, the greater her fear a major roadblock to a quick mission lay ahead. She fought back rising concern and focused on the grasses brushing at her legs. Even though they stayed close to the fence to hide their trail, big cats hid in this environment, waiting to pounce on unsuspecting or distracted prey. Taljaard's revelations definitely had *her* distracted.

Twenty minutes later, they encountered a protective canopy and moved along established animal paths to another gate. A short hike later, they stopped behind trees near a farmhouse. As she wiped her forehead on her pink shirttail, Taljaard handed over a water bottle.

"You redhead types ought to wear bigger hats." He tipped his wide-brimmed khaki on but didn't offer it to her. "Zimbabwe has only two seasons. Hot and damn hot. American Girl Scouts are supposed to be prepared."

"That's the Boy Scouts' motto." She finished off her water. "Besides, I hadn't planned on a nature hike."

Taljaard turned away and observed the farmhouse through binoculars he had produced from his pack. The rifle hanging over his shoulder spoke volumes about his opinion of possible trouble ahead. Five minutes later, they moved to a different position, stopping once again to inspect the farmhouse and its surroundings. After multiple reconnaissance stops, she had seen the brick structure from every possible angle.

The red tile roof, oversize windows, and flower beds hinted at a prosperous not-so-distant past. Today, though, a wide cement porch with a green-painted floor had only a single chair sitting by the front screen door. The white wood trim had dulled with dust. A trellis hung askew, pulled loose under the weight of an untamed bougainvillea. Weeds choked a row of red hibiscus bushes.

So far, Taljaard had remained alert but relaxed. That boded well for a quick retrieval of the material, if indeed this was their destination.

At the rear of the house, a young black boy, barefoot and dusty, kicked a well-worn soccer ball across the yard. Dimples appeared on Taljaard's cheeks as he pocketed the binoculars. "Mrs. Lubbe has tea and lemonade waiting."

Her mouth had barely opened to protest the waste of time when he cut her off. "She believes you're a friend here to help. Call me Ian and take time to socialize."

"But we don't have—"

"The information is in the hands of someone whose trust you

must gain. Every move you make will be examined. It will take time."

"Time is the one thing we don't have. Damn it, Taljaard—"

"Ian."

"I didn't hike half a continent for lemonade. You coerced me into coming here because you don't actually have the information, do you?" What remaining fingernails she had dug into her palms. "Lemmon must have known this all along."

Taljaard—correction, Ian—ignored her and waited until the boy disappeared inside before he walked toward the house. She had hoped this was the end of the information trail. Now, her intuition feared the journey had just begun.

Ian, as she forced herself to think of him to prevent any unintentional slips of the tongue, had the boy scooped up in his arms and locked in a bear hug by the time she slipped in the back door.

"Here's the surprise I promised." Ian turned at the sound of her arrival.

The boy's face lit up with joy as he extracted himself from Ian.

She removed her sunglasses and slid them into a pocket on her shorts. Ian patted the boy's shoulder. "Sipho, meet Miss Joni."

See po, she repeated in her head, and stuck out a hand.

Ian encouraged him to greet her. "*Sawubona.* I'm eight." With his chin dipped shyly, he took Joni's hand. For such a slight figure, he shook with gusto. "I like Ian's friends."

"Nice to meet you, Sipho. Is this your house?" She made a stab at getting information even if she had to stoop to questioning a kid.

"No, I visit Mrs. Lubbe. Do you play games?"

Was this some strange test or a delay? She glanced at Taljaard, who shrugged. It had been a few years since she'd played with her kindergarten niece and nephew. What *quick* games did eight-year-old Zimbabweans play?

"Sure. Do you have Uno? Chutes and Ladders? Battleship?" That about exhausted anything she remembered.

Sipho shook his head. "I have chess."

Of all the things she expected on this mission, a chess game with a confident eight-year-old Zimbabwean was the last. Ian offered her an imperceptible nod of encouragement.

"Be ready for a challenge," she said. "I play to win."

Energized, Sipho bounded down the hall, nearly wiping out an elderly white woman balancing a tray of cool drinks. Their visit was expected.

The lady, who could have passed for her grandmother, laughed and called out, "The chess game is on the kitchen table, dear."

"Afty, Mrs. Lubbe." Ian removed his hat and put it on a hook by the back door.

"Closer to teatime, I'd say." Mrs. Lubbe gave Ian an affectionate wink. "In here." She led them into the parlor, where she plopped the drinks on a round coffee table. "Make yourself at home. Didn't hear you drive up. A bit of trouble?"

"Nothing I couldn't handle. Brought a friend with me. Meet Miss Joni."

"Refreshing to see a pretty young face in these parts. Your first time to Zimbabwe?"

"I've seen the falls from the air." She shook Mrs. Lubbe's callused hand. "Do you live here alone?" Was someone else in the house, gathering the information for Joni to take?

"My husband has gone to town. Our son moved to Zambia to farm." Moisture filled Mrs. Lubbe's eyes. "Afraid for his wife and child's safety, he was."

The real emotion suggested Mrs. Lubbe wasn't an active player in Taljaard's network. "You surely must miss them. Might I ask why you stayed on?"

"This has been our home for three generations. Used to have more help, but can't afford to pay them now." She handed Joni a lemonade. "I suspect Ian told you we lost everything. Our land was redistributed to three proprietors. Only one, a major in the army, showed any interest in ranching. He offered part of the proceeds if we'd stay and raise cattle for him. It's hard not to hope these bad times will improve. At the moment, we expect to see little compensation."

"Is the boy the major's?" Joni asked.

Mrs. Lubbe glanced to Ian and back. "Heavens, no. We hardly see the major. He lives in Harare."

"I'm ready, Miss Joni." Sipho sat on an overstuffed couch with the board next to him and pieces arranged. For all the turmoil around, he seemed a happy child.

Joni stayed in place, ready to snatch the info and run. She hated to disappoint the child, but duty called.

"Be ready for a challenge, Sipho." Ian nudged her toward the couch. "Miss Joni plays to win." He whispered in her ear, "I need prep time."

Tired of being manipulated, she sat on the couch, nearly bouncing the pieces off the board. She caught the bishop, a hand-carved wooden figure, carrying a long shield and spear like an old Zulu warrior. The rooks of the intricate set resembled circular thatched huts. In the place of her king, though, sat a polished granite stone.

"Mrs. Lubbe lost the king when some men messed up her house." Regret played on Sipho's face. "It's okay, because you won't move him much."

Mrs. Lubbe, sitting in an overstuffed chair, cleared her throat.

Sipho glanced at her and fingered a pawn, biting his lower lip, apparently deep in conflicting thought. Slowly he reached out and turned the board around. "I like the rock king. You can have my men."

Sipho's charm ate into her heart. Whether American or African, kids wanted the best, the prettiest, or the biggest. She spun the board back. "I plan on winning so fast I'll never even touch the king."

A broad grin spilled onto his face. "I move first."

Ian drifted off to the kitchen. How long would it take for him to prep the package? Why didn't he already have it ready for travel?

Sipho made simple errors in the early moves. Tempted to checkmate him with a king's gambit and get on to the job of collecting the information, she fingered her bishop and stared at his innocent face. *Oh, hell.* She moved a pawn instead.

His lower lip pressed into a frown as he focused intently on nothing but the game. With passing minutes, more of his pieces than hers lined the edge of the board. Where the hell was Ian?

Bang!

The parlor echoed in sound. On pure instinct, she tackled Sipho to the floor, hiding him under her body as Ian appeared and launched to a window with rifle in hand.

Mrs. Lubbe giggled at the sight. "That's Farai, one of the major's employees, on his motorbike. Fuel is in short supply. Who knows what he puts in that engine."

Breathless and with adrenaline surging through her veins, Joni stood and offered Sipho a hand. He tightened his arms about his head and curled into a fetal ball. She touched his arm. "It's okay, sweetie. The bang was just a motorbike."

"No." He pulled away from her touch.

Surprised at the vehemence of his rejection and that her caring touch had zero impact, she looked helplessly at Mrs. Lubbe, who sadly shook her head. "Give him a few minutes. He'll come around."

Believing Mrs. Lubbe knew best and wondering what caused this strong emotion, Joni picked up the chessboard lying on the floor and set it back on the couch. Children shouldn't have to live this way. For the first time, she felt empathy with parents who lived in crime-ridden areas.

The rock king lay at her feet, but the lighter chess pieces had scattered across the floor. As Ian remained at the window, she collected the men one by one and set them back on the board.

"I guess we'll have to call this game a draw. I've no idea where all the pieces go."

"I do." The boastful claim came from under the thin arms covering Sipho's head.

"There are thirty-two men in the set, half of which we'd moved or had been taken."

"I remember." This time he peeked out and nodded with assurance.

"I'd stay and play, but have to get going." Precious daylight slipped away.

Ian stepped back from the window. "You've time."

His eyes told her not to question his pronouncement. What was Ian waiting for? She picked up a chess piece. "Okay, let's see how close we can reset the board." Sipho, still curled tight, didn't respond. "Your bishop was on G seven."

"What's G seven?" Sipho dropped both arms to the floor and frowned in her direction.

"The length of the board is numbered from one to eight, and across the board from *A* to *H*. So if I go up to seven and over to *G*, that's the square I want."

Sipho rose to his knees and looked over the board. She explained the pattern one more time. "My bishop wasn't there. It was here." He took it from her hand and then proceeded to accurately replace the pieces—all but one, which now protected the queen she had threatened.

"How did you remember all this?"

"I remember lots of things. Ian knows. We play games."

"What kind of games?"

"He sets the board up and puts the pieces in weird places. After I see it, he cleans off the board. I remember where the pieces went."

"Do you want to play that now?"

"No." He gave her a confident grin. "I want to beat you."

Ian once again disappeared. For the next ten minutes, Joni did her best not to overpower Sipho, but she was determined not to hand him the game, especially after his rearrangement to protect his queen. His youthful face creased in concentration as he contemplated a move to check her king.

The screen door slammed behind Ian. "Sipho." His stern tone alerted Joni to trouble. "Go get your stuff. Quickly, like we talked about."

"I want to win."

"Now."

When the command produced no action, Ian knelt before Sipho. "We don't have time to argue, buddy. Miss Joni is through with the game, too. If you want to help me like we talked about, then it has to be now."

Sipho turned his head away.

Joni took Sipho's hand, but he pulled it back to his lap. "I promise you another game after you do what Ian asks. I may even teach you a move guaranteed to beat him the next time you play."

Her offer weakened Sipho's resolve, and he slid off the couch. When he disappeared, both she and Mrs. Lubbe shot to their feet.

"There's a police truck coming fast down the front road." Ian kept his pronouncement low so Sipho couldn't hear.

"How do you know?" Joni asked.

"I climbed the blasted front tree."

She looked from Ian to Mrs. Lubbe. "We're out of time. We have to go."

Mrs. Lubbe nodded, grabbed the glassware, and slipped away.

Ian hustled Joni to the back door. "We must move fast. These men are dangerous and won't hesitate to shoot. Whatever you do, make Sipho think it's a game or he may refuse to move."

A lump stuck in her throat. "Will Mrs. Lubbe and Sipho be all right?"

"Mrs. Lubbe will be fine. She'll head to a neighbor's."

Joni heard the elderly lady hustling around the kitchen. "Does she need our help collecting anything?" Like the information?

"She'll be happier if we stay out of the way." He took a few quick steps down the hall and called out, "Sipho, you're as slow as a bat on a sticky wicket. Hurry."

Out of breath, Mrs. Lubbe darted from the kitchen toward Ian with a cloth bag clutched in her hand. "It's not much, but it should hold you over." She handed the bag to him. "Sipho eats like a lion."

Ian gave her a light hug and kissed her cheek. "It's perfect. Off with you now."

Mrs. Lubbe stepped past them out the back door. One last look at Ian showed tears filling her eyes. "Take care of Sipho. He'll need you close at night." She let the screen door shut.

When Mrs. Lubbe clunked down the steps in her sturdy shoes, the ugly truth hit Joni. She snatched the bag from Ian's hands. Inside it held food.

"Where's the information?" She shoved the bag back at him.

"I have it."

"When? I never saw Mrs. Lubbe move from her chair after we arrived."

"Then you weren't very observant."

"Hell, Taljaard, you've had it all along. What's going on?"

Sipho walked up with a school pack and athletic shoes with no laces on his feet. "Don't be mad at Ian, Miss Joni. He promised to play hide-and-seek with me but said a man I don't know would pick time. I guess time is now. We can play chess when it's over."

Ian opened the door and hustled her and Sipho out, grabbing

his hat off the peg on the way. "Show Miss Joni to the rhino fence by the baobab tree. I'll meet you there." Ian slung his rifle onto his shoulder and started away.

Joni ran after him and grabbed his arm.

"You have to be crazy." Her fingers dug into his flesh as she dropped her voice so the boy couldn't hear. "Sipho can't come with us. We have to get that information back to Taz. Let him go with Mrs. Lubbe."

"He can't, and we can't go back to your plane with these men on our tail. We have to lose them first. It won't be easy."

"Of course it won't. We have a kid to keep safe."

"No, it won't be easy because we have you. Sipho knows this area well."

"I don't know what you're up to, Taljaard, but I've been your pawn long enough."

Ian caught her firmly by the shoulders, digging in his fingers to ensure her attention. "Look, we don't have time to argue. It's not my fault you weren't fully briefed and came in with preconceived ideas. We have to save the boy."

"Why, because he means something to you?"

"No, because he *is* the information."

Stunned, Joni stared through Ian, seeing nothing as her mind tried to assimilate his comment. "I'm rescuing a kid? Where are all the computer components or financial documents?"

"I told you, he's the information. While we stand here and argue, those men are coming closer. I'll explain later. Now follow Sipho to the fence."

He released her, but she grabbed his shirt before he could dash away. "Wait a minute. Taljaard. You don't get it. I don't *do* kids. I don't deal with them and they don't mess with me. I can't protect him."

Taljaard looked at her, puzzled, and, if she read it right, a bit disappointed, like she'd kicked a puppy or something. "You and Sipho did fine. Now I have to go or you'll both be dead."

He pried her fingers loose before she could argue further and raced toward the front of the house. His final touch had seemed stiff and unfriendly. Heck, she was trying to do what was best for Sipho. There was no way she'd admit to a stranger that the last kid she'd tried to rescue had died. For two years she couldn't even admit it to herself. If Lemmon knew about Sipho, then he'd played a brutal game with her head and heart.

"Come on, Miss Joni. This way," Sipho called. "It's part of the game."

Sipho had been too far away to overhear. His excitement showed in his spirited gait, forcing her to run to keep up with his youthful legs. Whatever game Ian led him to believe they were playing, Sipho wanted to win. And until she figured things out, so did she.

CHAPTER SEVEN

Joni rushed off after Sipho as he headed across the cropped grass toward the back edge of the yard. She and Ian had come in from the side, but Sipho appeared to have a different trail in mind.

A tethered goat stared at their antics and munched grass as Sipho disappeared behind some abandoned stables. She found him waiting there, grinning at the game challenge.

"What were the stables for?" She sucked air. Jogging was one thing. Sprints were another.

"Tourists ride to see animals." He didn't even sound winded. "No more tourists. Hurry." He set off again, light-footed and eager, full of youth and energy. She had lost hers following Ian into this out-of-control mess.

She shifted her focus from trying to maintain a low, stealthy profile during their escape to how easily she'd been manipulated by Obergen and Lemmon. Hell, not just by them, by the whole damn project office. She'd bet Sipho weighed close to the extra kilos added during the demo flight. They knew. All of them had to know.

But what were they thinking? She couldn't stuff Sipho under the seat or bolt him behind it. He'd have to sit in her lap. How could anyone even have remotely considered flying him out a wise idea? Surely Lemmon and the CIA could have spirited Sipho across the Zimbabwe-South Africa border in some other manner.

Her heart pounded harder. Not only from the exertion of running and dodging along uneven ground, but from thoughts of Obergen in bed with the CIA. Damn him for helping hatch such a stupid plan. His die-hard need for success would likely kill her and Sipho. Of course, the Americans probably ate up the idea of a stealthy, daytime operational test to push Taz's development. But to her, the entire scenario smacked of greed, power, and putting high and mighty achievement in front of the health and welfare of people.

She nearly fell over Sipho, who had stopped again under the wispy branches of a gnarled tree. Ahead the trail cut through a copse of small, gangly trees and thorny brush.

"Miss Joni. Your hair." He pointed at her and shook his head. "Way too red and shiny for game. You make it easy to find us."

Great, another critic in the wilds of Africa. "When I was a kid, we just found something to hide behind."

"Ian say hide like animals. Use environment."

"That's good advice."

"Ian teach me many things. He say I very smart." Sipho pressed his lips into a frown. "But need to practice more English. I speak Ndebele. Some Shona, too. I miss school."

Joni pulled back her hair and twisted it up under her cap. She hated to admit it, but the more she was around this kid, the more he grew on her. No wonder Ian had slipped under his spell.

But Sipho still didn't look happy with her appearance, and she could imagine why. Securing a handful of the dirt from the trail, she rubbed it into the pale skin of her neck and face. "Better?"

Sipho nodded, but shrugged as he indicated her arms and legs. "Still too white. But we must go."

She scooped another handful of dirt before following, rubbing it on her arms and legs as she trotted. Sweat helped it stick and smear. The camouflage might not be up to the military standards she had learned during survival and evasion training, but hey, they were only going to a rendezvous point, and the bad guys were back at the farmhouse. Still, if it helped Sipho think she was playing along with the game, then all the better.

"Do you and Ian play like this often?" Embarrassingly, her words huffed out. Exercise in an air-conditioned gym didn't account for the impact of heat and humidity.

"When I visit. And since I stay with him."

"Has that been long?"

The light pad of his athletic shoes ceased their steady rhythm. "Since *Ubaba* had to go away."

"*Ubaba?* Who is that?"

"My father."

A psychiatrist wasn't required to hear the worry at his father's absence. Surely his story related to her current quandary, but she didn't have the heart to dig further. Inquiring about his mother likely wouldn't be much brighter. Ian knew Sipho's story, though, and regardless of his wavering faith in her abilities, she expected answers from him ASAP.

His evasiveness about the so-called information and then having the balls to claim she endangered their escape pissed her off. Hypocrite. She might not be a save-the-day, super ass-kicking spy, but she sure as hell wasn't a liability.

"Hurry, Miss Joni. Not far now."

As the scrubby brush and trees faded into open savanna again, the rhino fencing she and Ian had traveled near came into view along with the majestic baobab tree.

With their rendezvous point in sight, and knowing Ian lagged well behind to give them time to escape, Joni was tempted to slow her pace. A low vibration changed her mind.

"Sipho, let me carry your pack."

"Why?"

"It will be easier for you to show me the tricks to win this game."

He grinned and handed it over.

The whirring of long blades cutting through air grew louder. Wary, she loosened the straps and slid his school pack over her shoulders. The opponent in Sipho's game had upped the stakes. A helicopter had joined the fray.

CHAPTER EIGHT

Ian doubled back to brush out Joni's tracks after checking on the approaching vehicle. Ruddy hell. The Americans and South Africans had their heads up their arses. He'd requested a bush-savvy operative. Instead they sent an argumentative, improperly briefed woman—a bloody redhead with an attitude, no less.

White men stood out enough in these parts, let alone a woman. And trying to pass her off as a farm wife was near laughable. Those women cooked, cleaned, plowed, and farmed. Their toughened hands and skin went well with hardened, solid bodies, softened by rounded middles, the result of hearty meals. Silken, pale skin with a light smattering of freckles and perfect curves that even baggy shorts couldn't hide meant Joni had fine, educated city woman written all over her. Not that he wouldn't want to keep such perfection to himself, but shit, the guy at the checkpoint nearly drooled to death over her.

At first Ian had tried to convince himself that the fact she was a woman might not play out so badly. Sipho had taken to her, and she offered a mother figure he needed. But then she spoke about him like he was a liability. She didn't *do* kids. What the hell did that mean? Lemmon should have sent a man.

Ian had to forget Joni and focus on the mission. Distraction killed. That's why women with their damn curves and fancy hair didn't belong in combat, and he knew what lay ahead today qualified as combat.

He sprinted to a brick shed in the side yard. The whitewashed wood door had been wrenched off in a prior farm raid and later loosely rehung when the major took over. Inside, a stripped workbench with the empty outline of tools told of the hardships facing everyone in the country. The Lubbes had lost much to looters, and yet the major had neither the understanding nor the monetary inclination to restock or secure his recently acquired lands.

Only a simple hoe and shovel hung on one of a dozen now empty wooden pegs. Mrs. Lubbe did her best to maintain a vegetable garden, but obtaining seeds and irrigation water had become nearly impossible. He reached into a woven basket on the shelf and dug out a roll of duct tape, the universal tool.

In the far corner, amid empty green and red Castrol oil cans strewn on the floor, sat a drum barrel containing used motor oil. Ian swept up a rusty can and scooped it full of the fouled liquid. Brown oil dribbled down the metal and onto the dirt floor. He nested the can into a larger one and dashed to the front porch, where he secured the oil. A rumbling noise and dust cloud approached the turn for the farm. With only moments to spare, he reached his hiding place and waited for action.

A scratched and battered pickup barreled past the front fence line into the yard. Brakes screeched as the driver chose cooler temperatures over strategic position and stopped under the broad shade of a Boer-bean tree. Fool. Tree nectar wouldn't be the only thing dropping on this vehicle.

A blue stripe ran around the white truck body, and the word *police* stood out on the front hood. One glance said these guys hadn't come for a simple house call. He'd predicted more time before the pitiful gang he and Joni had encountered exposed their movements. His mistake. With one hand he fingered loose the knife strapped under his sock.

The driver and two men riding in the truck bed climbed out. All had tentative expressions and kept the vehicle between themselves and the house. Ian recognized one as Mzilikazi, named after a historic Ndebele clan chief. He had a wife and four kids, none older than Sipho. Ian smiled. Mzilikazi knew, and so did the others, they had to go through him to get to Sipho.

Logos on the men's matching blue caps showed they hailed from the nearest police station, a good forty minutes away. While poor local economics kept their pants and shirts from any conformity, their weapons were another story. Each wielded a Chinese-made AK-47. China traded military weapons and vehicles to Zimbabwe in exchange for economic advantages. Whether or not the policemen had chosen single or rapid fire, Ian couldn't tell.

The cab passenger whose face hid under a wide-brimmed circular hat disembarked last. One of the nervous policemen stepped forward as though to protect him. The local Ndebele didn't care for interference from a big-city official—and probably a Shona tribesman at that—but they respected the power this man wielded. The passenger reached out to grasp the door. An old scar slicing down his arm shouted recognition.

Kagona. Ian didn't have to see the man's face to know penetrating eyes assessed every aspect of the house. Ian's breathing quickened and his hands shook, not from fear of Kagona, but of losing control to the hate churning within. The presence of the top CIO operative on the farm meant someone in high government office wanted damage assessment and a quick resolution.

A mere week ago, Ian had sat across from Kagona in a CIO office outside Harare. If not for two armed bodyguards, Ian would have choked the life from the man's body. Burned into his memory was Kagona's arrogant voice and the smug tilt of his well-fed chin.

"Surely a white rancher like you cares little for the likes of an Ndebele boy?" Kagona had spoken as though his mere reputation for cruelty would make the truth flow from Ian's mouth.

"I'm just a wildlife manager. The government took my ranch."

"We're not here to discuss the pros and cons of politics. I'm more interested in where such a boy would go with his father gone and mother dead. You know his family. Who are their friends?"

Kagona, a Shona from the Zezuru tribe, had little love for the Ndebele or for other Shona not of his tribe. The Zezuru held several powerful positions in government. Kagona had no desire to see that change.

Ian had shrugged at his interrogator. "I'm the last friend of Sipho's family you've left standing. The others seem to have disappeared. Perhaps you've forgotten them in your jail or morgue?"

Kagona's fingers dug into his armrests. "Smart answers will only increase your chance of seeing our confinement facilities."

"I can't give you information I don't have." Ian had shoved to his feet. "Since there is nothing more we need to discuss, I've business before I return down country." Without any dismissal, he strode to the door barred by guards.

Kagona had taken a calculated risk and waved his guards back. All he had to do was tail Ian and eventually he would find Sipho. Kagona had been right about that, but Ian hadn't made it easy.

Ian swallowed hard in his concealed position, trying to think of the positives. Kagona's presence at the farm meant things were falling apart up in the capital and he had to stop Sipho's information from leaking out. Ian had to keep Sipho out of Kagona's hands. He owed his best mate, Nomathemba.

The situation chilled another degree as Kagona picked up a large black pistol from the truck bed. Ian recognized the extended barrel and polished wood handle. With several sharp, determined moves, Kagona breech loaded a disposable dart and pumped up the air pressure. The bastard desired—no, needed—the boy alive. The bosses in Harare must want info on whom Sipho had talked to and what he had told them.

Ian riled at the thought of a dart embedded in Sipho's flesh. The wrong dose of sedative could kill him.

Kagona signaled his men close. "I want Taljaard and the pilot alive. I need to find that aircraft. Then bring me the boy."

Well, well, how in bloody hell did Kagona know about Joni's arrival? Suspicions raised concerns. He'd sent Sipho off with a stranger who might be complicit in whatever scheme Kagona had enacted. Damn. He had to clean up here and catch up with them.

Mzilikazi was assigned to secure the front of the house and guard the vehicle, while another policeman moved to the secure the rear perimeter. The third man preceded his leader to the door, AK-47 ready. They didn't bother to knock.

In the distance, the steady beat of air said a chopper approached. One more obstacle to overcome, but he'd handle the hurdles one at a time.

As soon as the screen door banged shut behind the intruders, Ian swung from the Boer-bean tree and dropped onto Mzilikazi below. The impact of Ian's boots against the man's shoulders dropped him to the ground with a thud. A quick check showed Mzilikazi was unconscious.

"Sorry, buddy." He squeezed the poor guy's shoulder. *One down. Three to go.*

"Run," Joni yelled. Sipho's long, dark legs blurred under him as he ran across the open space toward the baobab tree. If they were

72

lucky, the helicopter would head for the farmhouse first and not notice them in the distance.

A chopper appeared on the horizon above distant treetops. Luck wasn't with them. Unease spread through her. One glance at the offending machine pegged it as a tadpole-shaped Alouette III, a military-type helicopter flown by the Air Force of Zimbabwe. Only this one had no markings, no tail numbers, nada, not even advertising for a Victoria Falls tour or safari.

Someone had power to utilize a helo to recover a mere boy. Out of a good one hundred-plus AFZ aircraft, the briefing had reported, only twenty were operational due to embargoes, lack of spare parts, servicing requirements, and expensive aviation fuel that had to be imported from South Africa. Civilians in these parts didn't fly Alouettes. The info in Sipho's head must concern a person high up in the government pecking order. Lemmon and the CIA probably knew that, too.

She and Sipho had nowhere to go but forward toward the baobab. The rhino fencing ran off to their right and the chopper closed in from the left. She kept her eyes fixed on the craft as they ran. It approached with the nose and tail a bit skewed along the line of flight. Years of flying gave her good reason to guess the pilot must be short on training hours. He crabbed like a damn novice.

The crew door slid open as it passed overhead for high recon. Blasts of air from rotor wash kicked up dust and dried grass. Sipho shielded his eyes and slowed down. As the helicopter moved away, he looked up, disoriented by the winds and noise.

"Keep moving. They're coming back." Her voice had risen from a yell to a scream above the din.

His glazed expression left her fearing he'd curl up and withdraw again. At all costs they didn't want to stop like frightened animals when the chopper hovered above. Movement meant safety.

She retrieved a soft cloth from her pocket, normally used to clean sunglasses. A quick tug left her with two soft pieces. She scrunched one up.

"Sipho, this will make the noise quieter." Mesmerized watching the chopper, he let her stuff a piece in each ear.

Done, she stepped between him and the chopper to break his focus. "Come on, we can beat them out." They moved again toward the baobab.

A man dangled his legs from the Alouette's door as it swung back over for a low pass. He leveled a rifle at them. Ghastly scenarios played in Joni's head. She shoved aside thoughts and went into action, prodding Sipho ahead.

The pilot slowed and hovered a mere twenty feet above the ground, driving them toward the fence. With a good thousand feet to the baobab tree, they'd soon be ducks in a shooting gallery.

The increasing rotor wash rushed down and out from the chopper, pushing them along and sending grit airborne. She grabbed Sipho by the hand and cut away from the fence. The air blast knocked him over.

She steadied him on his feet and yelled over the deafening noise. "Zigzag to the tree."

Sipho covered his ears and shook, simply trying to keep his meager frame upright under the increasing wash. She did her best to shield and push him along. The helo pilot worked to orient the crew door to face them, but the craft drifted away—another result of poor training. The pilot struggled, and she planned to make his problems worse.

As the chopper crept back toward them, something thudded to the ground near Sipho. Confused, he grabbed onto her.

"Don't stop, Sipho. We almost have them beat. Get to the tree."

They made another hundred feet before another missed shot plugged the ground. They weren't shooting bullets, but darts.

With the pilot's unsteady controls and their erratic movement, picking off Sipho with darts would be like hitting a bounding antelope—tough and near impossible.

Sipho turned, and with his arm shielding his face, looked back at the helicopter. He pointed and said something she couldn't hear. Her hat blew off and hair whipped against her eyes. She prodded him on, but he pulled her down to his face.

"Net, Miss Joni."

It took a second for the word to register. The guy in the door had exchanged his dart gun for a rifle with a large cone-shaped flare on the end. A net projector.

She scooped up Sipho by the armpits and half dragged him toward the chopper tail. If this guy flew as badly as she thought, the hardest thing for him to do would be to back up or turn 180 degrees in place. Before the man could fire, they found temporary shelter underneath the chopper.

She huddled over Sipho and folded her hands like a megaphone around his ear. "Be ready to run again. We're going to win this game." The shaking she felt under her hands said he didn't seem convinced. She wrapped her arms around him and tried to shelter him from flying debris.

Seconds ticked by as the wash beat down on them and her tension pulsed with each heartbeat. The rotors kicked up power and the copter lunged forward, moving away and leaving them in cloud of dust.

"Run." She angled Sipho toward the tree and the fence. They covered twice the distance they had lost as the chopper gained altitude and distance in order to turn around and come back.

Almost to the fence and fifty feet from the protective limbs of the baobab tree, the chopper caught up to them again. The projector fired, spiraling the net toward them. The spinning padded weights pulled outward, spreading open the mesh. It slammed into her back, knocking her and Sipho down. The impact stole her breath, but she rolled over, surprised at her freedom. The net only partially covered her body. The rest hung on the fence.

"*Akeli ncede!*" Sipho screamed, struggling in the net and further entangling his legs.

She shoved the net off her and crawled to her feet, yanking the nylon mesh with her and pulling it off Sipho. The darts would start flying any second. Terrified, Sipho curled his arms around her neck and lassoed his legs about her waist. She tried to run, but even as slight as he was, there was no way she could carry him far.

The pilot struggled to keep the chopper away from the fence and the tree and again began drifting. She and Sipho had to take this opportunity to make it to the tree or the game was over.

"We've beaten them again, Sipho. The chopper missed us and is moving back." She pressed her cheek against his and did her best to sound confident. "But I can't run this way. You must help me."

Something about her commanding words or tone didn't sink in. He clung tighter. They only had seconds. "Ian wanted us to make it to the tree. I know I can do it. What about you? Do you think you can make it? Ian will catch up to us soon."

The little game player looked frightened, but nodded yes.

Sipho slipped from her hold to the ground and stood. Poor kid. None of this was his fault, yet he was paying the price. "Stay between the tree and the fence. The chopper can't reach us there." Hand in hand they ran to the tree.

The Alouette pilot had recovered and realized his prey was out of their weapons' reach behind the tree. Soon he would figure out the only way to reach them was on foot. The chopper hovered to the far side of the baobab, away from the fence.

The tree and distance absorbed some of the terrific sound, giving them a moment's relief. Their options for escape, though, had dwindled to none. They were trapped. "Sipho, I'm out of ideas. The tree won't work as a hiding place once they're on foot. Ian won't be here in time."

"I know good place to hide."

"We can't run back now, it's too far. They'll catch us."

"No, Ian and I play in this place." Sipho pointed on the other side of the fence into the animal reserve.

"Do you know how to get through the fence?"

"Yebo."

"You know the combination to the gate?"

A confident smile broke through the strain on his face. "Ian teach me." Those words were fast becoming Sipho's favorite phrase.

Ian might balk at involving the precious conservancy, but it offered their only hope. Joni swept Sipho into a hug. "Come on. We'll beat these guys yet."

Chopper pilots kept their victims in sight. Even a poor pilot must have mastered a few of the basics and practiced this rule. She moved to the perfect spot where the Alouette pilot could land and still see her. What he didn't realize was the giant tree blocked his view of his prey's escape route—the gate.

A crewman hung out the chopper door, checking the remote landing site for stumps, holes, and other obstacles. If the pilot had any brains, he'd make one more sweep. Of course, considering the rest of his poor flying antics, she didn't count on that.

"Get ready to run when I tell you and open the lock," she said to Sipho. All she had to do now was wait until the pilot committed to touchdown.

The pilot plunked the wheels to the ground, knocking the crewman off his feet. A gunner's belt jerked the man back, and he fought it for the quick release.

"Go. Open the gate."

In a rally worthy of a great game player, Sipho raced off. Joni waited, allowing the crewman a few more seconds to focus on her and believe them trapped. Her cue to run came as he jumped out with the dart gun in hand, followed by a friend with an assault rifle.

The tree trunk hid the approaching men as Joni joined Sipho. He punched away at the cipher lock as the rotors slowed to a resounding *whop, whop, whop.*

"It no working," Sipho cried.

"Try it again. You have time. Take it slow." Actually time had expired, but telling him that wouldn't help. The men hustled from behind the trunk as the click of the lock opening sounded.

"Go, go, go." She shoved Sipho through and slammed the gate shut, keeping her body between him and their pursuers. The men broke into a run and so did they.

Clumped wild grass grew tall enough on this side of the fence to reach Sipho's ears. It slowed their pace but offered better cover. She and Sipho headed for overgrown, willowy trees and scrub a good distance ahead. Behind them clangs echoed as the men butted their weapons against the lock.

At a shot and loud metallic ring, Joni glanced back. The man with the dart rifle aimed a pistol at the cipher lock. Guess he carried equipment for all occasions. The man with the AK-47 had headed back to the chopper.

The guy with the pistol fired again, and the gate swung open. He glanced up in her direction and hefted the pistol over his head in victory.

"Don't get cocky," she mumbled, "it's the last victory you'll get."

The rotors wound up again to a whir as the helicopter prepared to lift off. "They're coming again, Sipho, only this time there's a man on the ground and at least one in the air."

Sipho slowed to look.

"Keep going. We have to make that brush cover and figure out what to do next."

"I know what to do. We go to rocks and hide." He knew

enough not to point, but tilted his head toward some hills in the distance where rounded granite boulders and bare rock jutted out. "I know cave. You go to rocks. I make plane go away. I be like animal. Ian teach me."

While her brain tried to process the distance to the hills and what Sipho meant by playing like an animal, he dashed straight across the savanna instead of heading for the brush.

"Sipho, don't be crazy." The kid was sacrificing himself for her. It should be the other way around.

She took a dozen steps in his direction before he became hard to see. Reality hit. He had a better chance without her.

The victorious wise guy behind her closed in. His orders might be to bring Sipho back alive, but no telling what plans they had for her. She scrambled for the trees, however scraggly and gnarled they might be.

The deafening chopper once again hovered overhead, and she could see two men in the bubble cockpit. The man on the ground waved and pointed in the direction Sipho had gone. She used every second of his delay to increase the distance between them.

The pilot swept out deeper over the reserve, the very place Ian had hesitated to involve in this mess. He'd have to deal with that problem later. At the moment she had little choice of where to flee.

Ian could have prevented all this had he been up front about everything from the start. She would have waited at Taz, and when he arrived with Sipho, she would've suggested he bungee over crocodile-infested waters and been on her way. He must have foreseen that possibility. He wanted her to become acquainted with Sipho and bend her to his cause.

Well, he'd failed.

A chopper full of nasty men left her little time for bonding, and she'd be lucky not to leave behind dead bodies on his precious reserve. One of those bodies might well be hers, considering the man behind her had a weapon. She planned to address her weapon shortfall the next time she corralled Ian's sorry ass.

The beat of rotors swept closer again and then moved away, flying a search grid. The action buoyed her spirit and gave strength to her legs. Sipho had slipped their visual radar.

Way to go, kid. Now she had to do her part—and fast.

CHAPTER NINE

Ian pressed a piece of duct tape across Mzilikazi's mouth and performed a quick truss job. Twenty seconds later, he had spread motor oil across the green-painted concrete porch and created a slick by the door.

Hurried footsteps from the room-to-room search echoed through the open windows. Ian slipped back to the truck and, after a little rearranging of his trussed friend, shifted the gear to neutral and released the brake.

His plan had holes, but he'd adapt it to work. He needed time, a heapin' load of it, to get Sipho to safety.

Ian walked along with the truck door open, rolling it toward a corner of the house. At the optimal place, he set the emergency brake before lifting off his hat and ceremoniously placing it on Mzilikazi. The man, whom Ian had positioned in the driver's seat and taped to the steering wheel, stirred, becoming conscious as Ian reached across him for the ignition.

Mzilikazi's eyes grew round with fear.

"Hang in there, mate." Ian patted his back. "The worst part's over for you."

Ian cranked the engine, grinning as it made a screeching noise easily heard over a wide radius.

He dived for cover and slid along the brick wall of the house until he was concealed behind an overgrown hibiscus. Inside the house, footsteps pounded up the basement stairs as the perimeter guard yelled something in Ndebele about the truck and ran past Ian.

A swing of Ian's rifle butt to the poor guy's stomach and then head brought him down before he knew where the attack had originated.

Two down and two to go.

Ian slipped in the back door in time to hear a thud and see up the hall as the remaining policeman slid across the oiled-up front porch. The man screamed as his body dropped down the concrete front steps.

Kagona swayed on the doorstep, having stopped before hitting the oil. He recognized the danger both before, and likely behind, him. He attempted a turn about the time Ian wrapped an arm around his neck.

A seasoned fighter, Kagona pressed his chin to his neck to protect his windpipe. Without hesitation, Ian leveraged his arm across the side of Kagona's cheek. Using his other hand, he pressed the bony side of his forearm against the asshole's face. Kagona's jaw crunched and he screamed before slamming Ian back against a wall.

Plaster collapsed around them. The motion allowed Ian to slip his arm under Kagona's chin and lock it against the throat. He slipped a hand behind Kagona's head, pressing it forward. Pressure from both directions squeezed his enemy's windpipe to stop the flow of air.

Seconds later Kagona went limp in Ian's arms. Ian fought the desire to crush out the last slimy vestige of life. A vision of Nomathemba's round, soft eyes and wide, caring smile chose this inopportune moment to interfere. That wasn't her way or his.

He released the pressure and let the unconscious Kagona

drop hard. He picked up Kagona's dart pistol and leveled it at the oil-streaked guard, who moaned and struggled to his feet.

The guy stuck up his hands and stood, his hat askew and scrawny body shaking with fear. These men had no business doing the CIO's dirty work. They had their hands full enough with local squabbles.

Ian shoved Kagona across the slick porch and picked up the man's dropped AK-47. "Drag him," he commanded.

The policeman tried to lift Kagona but gave up and dragged him down the stairs and through the dirt and grass to the corner of the house.

The perimeter guard sat in the dirt clutching his head.

"Help your friend." A gun did wonders in obtaining quick cooperation. Ian switched back to his rifle, then collected the last of the AK-47s from behind a bush where he'd tossed them in his haste to get in the back door. He dumped the ammo in the truck before tossing the AK-47s and dart gun on the ground in front of the pickup.

"Over there." Ian pointed to a small fenced enclosure. The men's gaze remained glued to Ian and his rifle as they dragged Kagona over.

Ian pulled back the flimsy fencing, exposing a wide hole in the ground edged by crumbling bricks. He remembered when he was a child and the old dug well actually held water. When drilled wells became common, the dug well went unused except for as an occasional clean waste dump—the perfect place for Kagona.

A piece of sun-bleached, weatherworn rope hung on a rusting metal post. "Down there." He pointed to Kagona and then to the well.

Without complaint, the men nodded and tied the rope around Kagona. His descent into the well wouldn't be graceful, but it was more humane than dumping him in.

One held the rope while the other slid Kagona up over the

dirt and the disintegrating brick ledge. The policemen struggled to hold the dead weight as they lowered away. The rotting rope started to shred. Their fearless leader awoke, sending a muffled scream echoing from the well. The rope twisted and snapped. A crunching thud followed by cursing expressed how Kagona was enjoying his new quarters.

At gunpoint, Ian prodded the men behind the shed to where a wood ladder lay propped against a wall. They carried it to the well, where he extracted a rope from his pack and tied it to the uppermost rung. The men stuck it into the well and watched as it disappeared.

Ian pointed at the policeman who had played slip and slide off the porch. "Down."

The bruised man frowned, probably not eager to join Kagona, but followed orders. He slid down the rope until he reached the ladder and then climbed to the bottom. The second policeman followed. Ian hauled the ladder partially out before getting Mzilikazi, who was still taped to the steering wheel.

"You won't be needing this anymore." Ian took his hat back. "Now, join your friends."

Having watched his buddies from a front-row seat, Mzilikazi knew what to do. He scurried to the well and lowered the ladder, not bothering to pause as he disappeared over the edge.

Ian hauled out the ladder and retrieved his rope. "Don't worry, fellas. Farai, the major's man, will be back. If he's sober, he might figure out where you're hidden. By the way, nice of you boys to bring me transportation."

"You're dead, Taljaard." The words, jumbled from Kagona's injured mouth, rose from hell.

Ian gave a haughty salute to his captives and tried to laugh at the circumstances. He found it impossible. A helo circled nearby, and he imagined it followed Sipho. Sending Joni with him had been a mistake. The boy had a better chance on his own.

Ian gunned the police pickup and drove over the weapons stash a few times before heading off toward the action.

Dry, thorny brush grabbed at Joni's clothes and scratched her legs as she barreled into the scrubby area. Animal pathways existed, but staying on them made her easy to track. Off them, the going slowed and noise increased as debris from dead material clogged passages.

Evasion lessons learned at survival training in the woods of Washington State gave her ideas to work with. These thugs likely had no such expertise, and she hoped to use that to her advantage.

Although her pursuer had a weapon, she wasn't unarmed. Nature, and in particular the reserve, offered possibilities. Everything in her surroundings was a tool, both natural and man-made, and the last few minutes with the chopper had done wonders to improve her camouflage. Dust had left her dark blue shorts a dirty brownish-blue and taken the sheen off her pale skin. She missed her cap, but discovered reds and oranges from fruits and flowers helped disguise her hair.

Up ahead a dead tree had toppled not five feet from the rhino fence. It was a perfect place to hide. Joni smiled and headed straight for it, lengthening her strides, sidestepping, and trying to cover her footprints. Along the way she tore at small green or dried pieces and poked things into her waistband, collar, and hair. By the time she had settled into place, the sound of brush being shoved aside and dry litter being crushed underfoot grew closer.

Sunglasses hid the bright whites of her eyes and she used her ears to tell her when to strike. The noise and movement stopped. She heard a snicker and then a metallic click from a weapon. Did he have the dart gun in hand or the pistol? He probably had no plans to carry her out. Why should he when lions, hyenas, and vultures made the perfect cleanup team?

Right. Not over her dead body.

She resisted every urge to panic and flee.

Step by careful step, he approached the fallen tree. His quick breaths sounded inches away. A twig snapped. A knot grew in her stomach as she envisioned staring down a pistol barrel. His foul body odor constricted her throat.

His steps slowed and became controlled, yet moved past. From her prone position, she risked a peek. Sweat soaked through the back of his shirt as he stood a few yards away. His focus on the tree and an exposed part of Sipho's school pack had taken him past her position, where she hid under natural debris in a litter-filled depression. He paused an arm's reach from the fence at the upturned roots, his weapon ready to fire.

Joni bunched her feet up under her and launched toward him. He half turned at the explosion of noise, but had little recourse. He staggered under the solid branch she battered straight into his midsection. He flailed toward the fence, dropping the dart gun in an attempt to save himself from the unpleasant shock to come. Too late. Unable to stop his momentum, he closed his eyes and tensed awaiting the pain.

She expected an excruciating scream as his full body impacted the kinked wire. The fence accepted his weight as designed, caving beneath it but holding strong. Then nothing happened.

No sizzle. No zap. No nothing.

He opened his eyes and stared right at her, letting off a hearty, if not much relieved, laugh. His eyes narrowed as he glanced down at his weapon several feet away, near her. He pushed against the fence to right himself as Joni reached for the dart gun.

As her fingers curled around the grip, he pointed a pistol a foot from her face. *Shit.* He grinned and pulled the trigger.

A hollow click shoved death thoughts away. Guess there was a shortage of more than fuel in this country. She grinned back and with a quick upward motion planted the dart gun under the man's chin to wipe off his smirk. His teeth crunched and head snapped back.

She grasped the gun in the middle and whipped the barrel into the right side of his head, and then swung the butt around to the other. She finished with a solid jab to his solar plexus.

The man sagged back against the fence, sucking air and bleeding like a pig from the head shots. Electrified fence? Ian Taljaard was a lying bastard. She'd nearly been killed counting on that stupid fence. She yanked a leafy stick out of her hair.

"Lesson number one." She shook the stick at her hapless victim. "Never hide in obvious places. It's where people like you look." She hefted the empty pistol over the fence.

"Have a nice nap." She aimed the gun at his leg and fired a dart into his thigh before he even had a chance to cry out.

The attached holder had only one dart left. She strapped the gun over her shoulder, collected Sipho's pack, and hustled off to find the kid.

Joni knelt under the shade of a droopy tree. The rocky hills Sipho had pointed out were only a brief stretch of open grass away. Up to this point, a smattering of low trees with full canopies, thorny bushes, vines, and weedy wildflowers had offered enough cover for her to make headway each time the chopper swept away on its search for Sipho.

A black eagle soared high above, riding thermals and watching for movement representing its next meal. She related to the tiny creatures below, knowing a predator hovered above. The chopper broke off its search and landed back where they had entered the reserve. The pilot must have stopped to pick up the battered man she'd left in the thicket.

She sprinted the open stretch to the hills and squeezed into a weathered cleft in a granite outcrop before the chopper took off again. With a broad view of the flats below, she searched for movement. Sipho had to be down there somewhere.

She gave the chopper another fifteen minutes of gas guzzling before it had to land again and the men continued the hunt on foot. No matter who funded this excursion, aviation fuel was limited, and these guys were having little success from overhead.

The black eagle glided to a landing at a high rock aerie. Not a bad view from there. She could use more elevation herself to better pinpoint the enemy. Once again Joni waited for an opportunity before scrambling higher to a perch among colorful lichen-covered boulders. Two rainbow lizards darted by, giving her a start. Sheesh, she had to get a grip.

Near the foot of the hills, several antelope and two zebras grazed. They seemed at ease for such dire circumstances. A tap on her shoulder spun her around.

"Sipho!" Her heart thumped from the adrenaline his touch infused. Instinctively she wrapped him into a hug. "You scared me to death. How'd you get up here so fast?"

"Ian teach me."

Sipho had a serious case of hero worship. So much dust covered him, his rich brown skin had lightened to dirty beige. His button-up cotton shirt had lost all but one button and he had scrapes on his belly. He also looked tired.

"You're an amazing kid. You did become like the animals. Those men couldn't see you on the ground."

Joni and Sipho huddled down as the helicopter landed in a level spot near the edge of the hills. "I know you're getting tired, but they haven't given up. We still can't let them find us."

"I show you cave. We rest there."

"Unless the cave has a back door, it's probably not the best place to hide. These men may know about it, too."

He pulled on her hand. "It's secret. Come on. I take you."

Sipho seemed bound and determined to show her the cave. She owed him a few minutes of attention, considering all he had been through. Besides, moving farther into the hills couldn't hurt. The men were far enough away he'd have time to show her every nook and cranny and still leave them time to find a well-camouflaged not-so-*perfect* spot.

They hid their profiles by skirting well below the ridgeline and kept movement slow with numerous pauses behind rocks, bushes, or even trees whose roots dived into rocky crevices. Sipho's pace slowed more than necessary and his attention seemed to drift with a little cloud puff in the sky.

"Are we close?" she asked. He'd survived more trauma in one day than any adult should ever have to confront.

"Not far," he said slowly with effort. His energy, and hers, was running out. Enough dirt stuck to her arms and legs, she must weigh at least another five pounds, and a gallon of water to drink would be nice right about now.

A cleft in the hill exposed thick slabs of granite worn and separated along fracture lines. The weathering deposited crunchy granules of rock and sand along their path.

As she rounded a giant boulder that had fallen from above in past centuries, she lost sight of Sipho. She hurried to catch up, but he had disappeared. She stopped and listened. Nothing. Slowly she inched back the direction she had come. A long slit in the rocks she had passed was visible from this direction. It was tight, but a person could squeeze through it.

Sipho giggled in the dim light as she slipped inside the shelter of the cave. "You missed it." His voice echoed in the small chamber.

"My leader tricked me."

He sat on a cushion of dried leaves and smiled as he tucked his chin over his knees.

The filtered light exposed Bushman drawings. "Wow, I can see why a kid would like this. It's cozy, makes a great hideaway, and has animals on the walls."

"Mother and Ian play here, too. They show me cave."

His mother? Ian? Sipho had opened the door to Joni's curiosity, and she couldn't help herself. "Is your mother Nomathemba?"

Sipho lifted his chin to nod, and then buried his face against his bent legs. Perhaps now had not been the time to ask.

"As much as this cave is great, I think we need to go somewhere else in case the men come. There is no way out and we'd be trapped."

Sipho didn't move.

"Come on"—she held out a hand to help him up—"we've almost won this game. Don't give up now."

"Ian say cave safe. Stay here."

Joni was damn tired of hearing about Ian.

"Any other time it would be safe, but not when these men have guns. Come on, honey. We don't have far to go. You can lean on me." She gently grasped his arms at the shoulders and tried to coax him up. He simply clamped his arms tighter around his legs.

"We have to hurry and find another spot. You can hide like an animal again."

"No."

Exhaustion made his defiant tone sound more childish than his years. Most kids his age, no matter how tired, would count on an adult to help them. Sipho had seen the danger of these men. He knew this was no longer a game, and yet he had decided to block it all out and refuse to accept reality. He should at least trust her to do the best for him.

A memory tugged on her mind about the way he had shut down during their chess game, but she didn't have time to decipher it. The only thing that mattered was getting him out of here. Not an easy task if he chose not to cooperate.

Joni crouched and set a hand on his knee. He slapped it off without lifting his face from its buried posture. Things did not bode well for cooperation.

"Look, Sipho. We won't be able to see Ian if he is looking for us. Do think that is wise?"

Sipho didn't answer.

"I want to help you. Tell me what to do."

Still no answer. He rolled onto his side in a ball, locking his arms over his head as though shutting out the world. There was no way she could squeeze through the opening with him in tow.

Time for exploration had come and gone. She had to recon the enemy.

"Sipho, I'm going outside to keep an eye on those men. Stay in here until Ian comes. Promise me you won't go outside."

He remained silent and immobile.

Frustrated at her inability to connect when earlier they had seemed so close, she tucked the school pack behind his head. Unsure how well she could protect him against armed men, she squeezed out the entrance.

A kid had no damned business mixed up in this. Yet he had been dropped in her lap, or at least would be on the plane ride home. Thus, he fell under her protection, whether he wanted it or not.

A quick survey of the area near the cave uncovered an excellent defensible position. Joni crouched against leaf mold, soil, and natural litter wedged between a granite outcrop and boulder. The cave entrance was about twenty feet away in full sight and range of her dart gun. Below on the savanna, the chopper's crew door sat open. A figure with bandages wrapped around his head dangled his feet out the door. After a few seconds, he lay back across the chopper deck.

"That's what a little overconfidence can do for you," she muttered as the click of his pistol replayed in her head. She had wondered what people thought about before they died. For her it had been a refusal to accept death and a focus on her next move. She hadn't been able to get off a strike before he pulled the trigger, but afterward the redemption as she buried the gun in his flesh refreshed her spirit. Since when had she become so callous?

No one else moved on or near the chopper. Where were the pilot and the second man she had seen back at the baobab tree? She strained to see, but her eyes played tricks with the distance. What she would give for a decent pair of binoculars.

Down and off to her left, movement refocused her attention. The extra man she had seen onboard was barely two hundred feet below, carrying an AK-47, and climbing in her direction. So much for the secret cave and so much for dart guns. Guess he, like his compadres, didn't plan on carting her out, either.

A second movement appeared a mere hundred feet behind the gun-toting guy. The pilot? She caught sight of a khaki hat. Ian.

Relief tempered the dread that had been building when envisioning a lonely confrontation with two heavily armed men and she with a single dart. Perhaps now the odds would be even. Except she hadn't pinpointed the pilot's location.

Below and away from Ian and the first AK-47 man, she divided up the steep hillside and rocks into a search grid, assuming the men from the chopper had split up in order to conquer. Back and forth she let her gaze and head move, trying not be distracted by animals and birds. The pilot must be out there.

By the time she looked again toward Ian, he and his target had disappeared behind an indentation in the hillside. Pressing her face near the rough stone boulder at her position, Joni slid out along the rounded rock face for a better look. The surface changed from red and orange lichen colors to a streaky white. An odd odor filled her nostrils.

Inches away, a shadow moved at the same time she did. She froze nose to nose with a yellow-spotted dassie. The guinea pig-size creature watched her with curiosity. Shit—literally, or at least urine—accounted for the white streaks and odor. She had perched on top of their community.

The creature wrinkled its black nose as though assessing her threat. Joni remained still, afraid to even blink and send it running for cover with that deafening shrill scream. Her secured position would be history with the men so close.

An inch at a time, she backed up until the dassie wandered off. As Joni sat back and breathed a sigh of relief, she heard gravel and rocks rustle on the hillside below and a man grunt. Ian and his man must have had a scuffle. She peered out toward the sound. No one moved. Damn, she couldn't see what had happened.

She caught movement in the opposite direction from where Ian had been. The chopper pilot had heard the noise, too, and barreled up the slope. From the ease with which he moved and the thick arms carrying an AK-47, she bet he fought better than he flew.

He disappeared from sight. Whether he headed toward the scuffle below or toward her location and the cave entrance was unclear. Ian, if he survived, might not be aware of the pilot's presence, but warning him would give away her position. Sipho had to be protected at all costs.

If the pilot chose the cave, Joni had only one chance to bring him down. One lousy chance since the dart sedative gave him time to fill her with holes before he keeled over. At least if she hit the pilot, Sipho would have a chance to escape.

She'd never fired a dart gun from a distance. The weight of the dart would surely affect the trajectory. The pilot had to be close when she pulled the trigger. She had to strike first and successfully. Any hesitation in firing meant bullets for her.

Across from her position, enough room existed for cover after she shot. She had no way of knowing for sure whether the man she now heard approaching was Ian or the pilot.

Small rocks loosened by heavy boots cascaded down the hillside. Someone climbed quickly at a steep angle. A hundred feet and closing. It could be Ian, if he had taken out the two men.

Would he call out, or did he believe there might be others? What if she darted him?

That possibility produced a smirk. Not a bad way to get even for his failure to arm her, but no, she required his knowledge to get them out of this mess. She had no idea of Taz's location from her present position.

The footsteps ceased. Whoever approached had fought before and knew how to be quiet as he closed in for a kill. A slight crunch of granite gravel put him at fifty feet. Seconds ticked by in silence. He must have moved closer. Her palms sweat against the gas canister dart rifle. One shot. He had to be close, but she had to be sure.

A slight rustle sounded ten feet away. Could that be him? Her heart and breathing raced.

A high-pitched shrill filled the air as dassies tore across the boulders for shelter. Joni stepped out with the gun sighted and fired at a body plastered against the rocks spitting distance away. The dart embedded in the pilot's side.

She dived across to the protective rock as a smattering of bullets pinged off the granite, splintering and sending piercing shards of rock everywhere. One tore through her shirt and burned across her shoulder.

Shit, now she had no place to go. Supine in the dirt, she prayed for a fast-acting sedative. The man stepped into sight. The dart had fallen away, and he showed no signs of keeling over. He tucked the AK-47 barrel against her head. "Where is boy?"

"Out on the savanna where you lost him." Dirt clumped under her sweaty palms as they pressed against the ground, ready to move her body when necessary.

"Ha. He's in the cave."

"Sure. Go look." She sounded surprisingly steady for being only moments away from death. The pilot still showed no ill effects from the dart. Hadn't the sedative deployed? "Don't blame me when you can't find him."

The bastard smiled just like his buddy had done before he'd pulled the trigger. "I'll take the chance."

The pilot stepped back as though to get out of blood-spatter range. Perhaps luck came in pairs and the chamber would be empty. Sure, when giraffes flew.

She tossed a handful of dirt at his face and rolled aside as the gun fired.

CHAPTER TEN

The pilot's body jerked violently before he toppled aside. Blood oozed from his shoulder. Surprised, Joni checked her own body for holes before scrambling to grab his AK-47 and retreat back into her death-trap cubbyhole.

She peered out. Fifty feet away, Ian lowered his rifle. "Sorry, I didn't expect these bastards to bring a helo." He shouldered his weapon and dug something silvery from his pack. He tossed it in her direction. "Make sure he can't attack again."

Dazed and still trying to get her heart out of her throat and back into correct anatomical position, she stared at the roll. "Duct tape?"

"For securing poachers," Ian answered her unasked questioned before sliding into the cave. Good thing he didn't get emotional over having saved her life or blasting a hole in somebody. Hell, he didn't even ask about Sipho. Simply took off to find out for himself.

Her hands shook uncontrollably and showed no signs of stopping any time soon. If becoming an operative made people like Ian ice-cold in these situations, then she never planned to join the ranks. She rather liked emotions. Reminded her she was human.

The man blubbering on the ground and not yet under the sedative's spell brought at least one of those emotions alive—anger. She rolled the injured pilot facedown and put a knee in his back.

"You should have stuck with flying."

Hell, this man had planned to snuff her out like a cockroach. He screamed in pain as she yanked the arm of his injured shoulder back and bound his hands behind him. Perhaps she had been a tad judgmental of Ian. He probably found it easy to be calm when he wasn't the one eating the end of a gun barrel.

She stood and rested a boot on the pilot's butt. "Next time you plan to take someone out, crash your chopper on them. For you, that should be easy."

She slid the firing lever on the AK-47 to rapid fire and retook her original position. No saying whether men she hadn't accounted for still stalked these hills.

Minutes later Ian reappeared—alone.

Her heart threatened to move up her esophagus again. Had Sipho disappeared, or been captured, or had Ian no more luck in getting him out of the cave than she had?

"Where's Sipho?" she asked.

"He's worn out. Said he'd be out in a minute. You two must have had quite a lark."

"Oh, sure, it was a lark, all right. We were shot at and had a net dropped on us. Not to mentioned being herded by the chopper like animals to slaughter. What took you so long?" She wrapped her arms about the gun and found a non-white-streaked rock on which to sit.

Ian prodded the pilot with his boot. The man had succumbed to the sedative. Ian ripped off part of the man's shirt and then stuffed it against the wound. He yanked off the thug's belt and threaded it through the man's armpit and around his neck, cinching it tight to hold the cloth in place.

"Who are these guys?" she asked.

"Friends. The ones I said you didn't want to meet." Finished with his quick first aid, he took a position where both she and the cave entrance were within sight.

Ian bent his knee and propped a foot back against a vertical exposure of granite. He studied her appearance as though expecting it to reveal details she'd omitted of her latest adventures. Joni reached up and touched her dusty hair, frizzy with bush litter accents created by a rotor blow-dryer. Okay, so she looked a bit down and dirty.

He, on the other hand, had barely broken a sweat. The only dust on him hugged his sturdy veldschoen field shoes, and, unlike her, he'd managed to hang onto his hat. She and Sipho should have stayed at the farmhouse and let Ian take on the chopper.

"How did you outwit the helo?" he asked.

"Lousy pilot." She tipped her head at the snoozing perp. "That's him."

Ian grew almost thoughtful, an emotion she hadn't previously attributed to him. "I wouldn't have sent Sipho without me if I had known about the chopper. He doesn't like loud noises. He could've shut down anytime."

"What he didn't like were those men shooting darts at him. I don't know a kid alive who would."

Ian dug a water bottle from his pack and tossed it in her direction. As she caught it, he added, "Sipho's autistic."

Shit. Another complication. Little nuances she had noticed with Sipho finally made sense. One of her nieces had moderate autism. The girl barely spoke, though, and had much more pronounced symptoms. "And you planned to tell me this when— never? Right before takeoff?" She twisted the lid off the water in one swift motion.

Ian sketched something in the dirt with his boot. What else had he neglected to share?

"Sipho's high functioning," he offered. "Has something

people call Asperger's syndrome. Sort of a genius with numbers, memory, and things, yet a mess when it comes to relating to people. He can sing, but too many loud noises drive him nuts."

The water eased the parching in her throat. She wiped the back of her hand across wet lips. "He did fine with me. Do you have more of this?" She screwed the lid back on the neoprene bottle. "For Sipho?"

Ian looked at her askance, as if surprised she cared. "I left him a bottle." Ian caught hers as she tossed it back. He worked a weak smile on his face and shot her a gentle look, one asking for understanding and perhaps showing a touch of remorse for hiding the truth from her. He was good. She had to hand him that. Lemmon and the CIA could learn something from this guy.

"Yeah," Ian continued, "Sipho does better with adults than kids, but earning his trust usually takes a good bit."

"Don't underestimate him. He saved my butt by drawing the chopper out to the savanna. They never did find him."

"He has a sixth sense about how to blend in." Ian's chin rose and chest puffed as though his son was the star striker on a soccer team. "He's trusted you from the start. You're like his mum in some ways. Resourceful."

Yet doubt flickered in his baby blues. Evidently in other ways she didn't measure up. It might have had something to do with her announcement about not liking kids. It had come out all wrong.

"Is his mom your friend Nomathemba?"

"How did you know?"

"Sipho told me back in the cave. Where is she?"

Ian jammed the bottle back into his bag. "Dead. Killed by Kagona."

His voice cracked and throat visibly tightened. Now Joni understood his devotion to Sipho. He fought to save Nomathemba's son from the same fate. Hard to argue with him putting Sipho before her own measly demands to know the truth about this mission—although his lack of forthrightness had put them all at risk.

"How about Sipho's father? Is he Minister Edward Mukono, by chance?"

"One and the same. Kagona has him in prison."

She sensed Kagona was responsible for much of the misery in Ian and Sipho's lives and the reason for this mission. "Is Kagona one of the so-called friends you mentioned earlier?"

"The top officer in Zimbabwe's Central Intelligence Organization. He sees himself as the protector of his country by ensuring those in power stay in power. I left him in a well back at the farmhouse."

"Jeez." Her fingers unconsciously tightened around the gun. "Dead?"

"I was tempted, but simply cracked his jaw."

"Ouch. That'll make him grumpy. Why does he want Sipho?"

"To encourage his father's cooperation, among other things."

"I take it there's a hell of a lot more to that story?"

"There is, but no time to tell it now. I noticed there's one man left on the helo." Ian pointed a finger at his head. "Did you do the damage?"

Evidently he'd caught sight of the bandaged thug. "Yeah, no thanks to you. You claimed that damn fence was electrified. I counted on it."

"Sorry. Funds to keep it going ran out a while ago. With no guarantees the remaining conservancy owners will hang onto their land, no one is anteing up money to the animal cause. So I keep telling people the fence is hot, and few are willing to test my word."

"You could've told me."

Ian shrugged. "Habit."

"Habit, hell. Up until now, you haven't trusted me with anything. And don't feed me the need-to-know line. You could at least have given me a weapon." Emotions built over the past hour boiled over. "You risked Sipho's life. You risked mine—twice. At least Sipho said he'd meet me here and kept his word. He's a trustworthy kid, unlike his teacher."

"Don't get your panties in a twist." A damnable grin spread across his face as though oblivious to all that had happened. "If you even wear any. The truth is, I didn't have a weapon to spare. Besides, you proved a bit unstable with the dart gun. Fired like a giddy kipper before you knew who stood there. What if that had been me?"

"It wasn't. I knew."

"Bugger that. How would you know?"

"Your friends the dassies. You'd know better than to set them off."

"And if you were wrong?"

"Then you'd have to wonder what I did to you while you were out."

His mouth clamped shut, and he stood speechless. It rather suited him.

Joni brushed past Ian, surprised to feel a lusty surge as she did.

He firmly caught her upper arm and pulled her against his side. "What you did here saved Sipho. If that man had made it into the cave before me, he'd have captured Sipho, maybe hurt him."

His husky tone and surprising compliment skewed her emotions. "I'll do whatever it takes to get Sipho and his information out." She placed her hand on top of his. "Regardless of the obstacles in the way."

The appearance of a dimple said he caught her message.

"By the way," she added, not wanting to let him totally off the hook, "I expect a decent explanation of what he knows."

"Fair enough." Ian's grip loosened. "That's a nasty cut on your shoulder. I'll be glad to take a look."

She peeled his fingers off her arm, taking an extra second before letting them go. "Once we're far from these guys." She backed away, clutching the AK-47 for support, and headed for the cave entrance.

Night had fallen so long ago Joni fully expected morning to arrive any time. Sipho's secret cave had melted into distant memory as hours and miles had passed. Sipho rode on Ian's back while she carried the rifle and day pack with Sipho's treasures inside. The weight of both bored into her flesh and left every muscle in her upper body sore. Her shoulders thanked the decision to ditch the thugs' unreliable weapons and keep only Ian's.

They headed into a remote area near the conservancy where several Ndebele villages were situated. Her gait along the dusty back road held up surprisingly well, partially from the silly rhythms Sipho created with his songs.

"Deep in Zimbabwe, what will you find?" Sipho chanted as he rode on Ian's back.

"I'll find a *nkawu* scratching his behind." Ian took a little skip and scratched monkey-like under an arm, nearly dumping Sipho off.

"Okay, Miss Joni, your turn. Deep in Zimbabwe what will you meet?"

"I'll meet a *mvubu* with four big feet."

Sipho laughed. The thought of her and a hippo nose to nose *was* a bit silly. She had sung this song at least ten dozen times tonight and now knew the name in Ndebele for any animal moving in the country.

Ian studied her exchanges more closely with Sipho. He must be quite confused. Hours earlier she had proclaimed she didn't deal with kids. Now she playfully returned Sipho's attentions. To be honest, she hadn't spent five minutes around any child for two years. She didn't want them counting on her as an adult, expecting anything of her she couldn't provide. Her brother and sister wondered why she wouldn't visit more. She loved her nieces and nephews but simply couldn't handle the reminder.

Yet, try as she might, she couldn't ignore Sipho. He'd done a decent job of getting under her skin. Sipho caught her studying him. So had his hero.

"Look." Sipho pointed ahead.

A reddish glow and burning cinders rose in the distant night sky. Civilization at last. Famished and fed up with trekking the countryside when she had a perfectly good plane to fly, Joni still doubted Ian's logic with this move.

"Tell me this is where we stop. We've ditched a police car, evaded a roadblock, and trekked God only knows how many miles along this rutted road—"

"Eight kilometers," Ian interjected.

"Only eight, eh? Well, I'm done. Personally, I think we should have headed straight back to Taz."

"It wasn't safe and you couldn't take off at night anyway." Ian shifted Sipho on his back. "Kagona has people watching all the conservancy areas and old ranches I've been known to frequent. Your plane is parked on one of those. I didn't want to point it out for him. He'll assume I'd be out of the area by tonight."

"You give yourself quite a bit of credit. If I were Kagona, I'd assume you'd be stuck without transportation and quite noticeable among the black majority."

"You would've been wrong. He knows my capabilities."

Kagona might, but she damn well didn't. Ian's dossier had said precious little about talents above and beyond being a rancher and wildlife manager. She'd seen him in action, though. He'd learned a few tricks somewhere, and it hadn't been from the animals or even poachers.

"Tomorrow," Ian continued, "instead of leaving town, we'll

be coming back. You'll take off in Taz with Sipho, and this Zimbabwean adventure will be no more than a bad memory."

Sipho gripped tighter around Ian's neck. "Who is Taz?"

"A friend of Miss Joni's who brought her to Zimbabwe. You'll meet him tomorrow. He'll be quite a surprise."

"Tell me," Sipho demanded.

"Is a surprise as fun if you already know what it is?" Joni asked.

"Yes. Well, maybe." Then he quieted for a moment and honestly said, "No."

She patted him on the back. "I want meeting Taz to be special. So you'll have to wait till tomorrow."

Sipho had held up pretty well. They had stopped for a rest and he had managed to sleep, confident and secure in Ian's arms. A brief snooze on one of Ian's broad shoulders sounded good to her, too.

Ian had doled out snacks to Sipho from Mrs. Lubbe's lunch bag but neither offered her any food nor took any for himself. He'd also given Sipho a bunch of thin, colored leather strips. When the boy wasn't sleeping, he'd been weaving the strips into some sort of long, square braid. The task, some kind of fine motor skill therapy, seemed difficult for him, but not impossible.

Joni slung Ian's rifle to her opposite shoulder for the fortieth time, wincing as it rubbed her still-untended cut from the flying rock. "Won't our presence put this village in danger?"

"Possibly, but they deal with threats every day."

Melodic voices floated on the night air. The beautiful, gentle strains of perfect harmony were not the beating drums and pounding feet she expected to hear in an African village.

"Not exactly a low-profile time to intrude." She noted the voices all were male. "It sounds like a party."

"It is. An Ndebele chief for the villages in this area was inducted this afternoon. We missed the formal ceremony."

"Good thing. I had nothing to wear." What she could see of her dust-covered shorts and shirt was creased and wrinkled. Blood spattered her pink shirt from what she hoped was the cut on her shoulder and not the plugged pilot. She looked like shit and rather smelled like it, too. "Do we really want to make a public appearance?"

"This is a private celebration."

"With a whole neighborhood invited. I'd classify anything more than a friend or two as dangerous."

"This is Ndebele territory. They're not the tribe in political power in Zimbabwe. That is to our advantage. The chiefs are sanctioned by the Shona government and allowed to run their villages according to tribal customs. As long as no Shona laws are broken, it works. Whatever spies the Shona had to check on compliance left after the ceremony. Kagona and his men won't be here."

"Sounds a bit naive. Slap a few dollars on a palm and anybody can become an informant. In case you haven't noticed, people in this country are starving."

"Sipho's mother came from this tribe. They have as much to risk as we do."

Sipho readjusted his hold around Ian's neck. Hearing anything about his parents must be painful. She and Ian needed private time without youthful ears overhearing so she could ask about Sipho, his information, and his family.

"What does the tribe have at risk?" she asked.

"Recent laws require a permit for more than three people to meet. The induction ceremony today had a permit. I doubt this celebration does."

"Three people? That's ridiculous. How can anyone get anything done?"

"Ah, you're already starting to understand some of their challenges."

Heck, half of Africa had major problems, and at the moment

she had no desire to solve them. Her feet hurt, stomach growled, and eyelids drooped. She'd been up since four a.m.

Shadows of dwellings stood out against the backdrop of campfire light. Some were traditional round clay and wood huts with thatched roofs, but several had plastered concrete-block walls with painted geometric designs. "What is this place?"

"A kraal. It's a family living cluster. Chief Zwide's family lives here. The last chief died a few months ago. His first wife still lives in the biggest painted house over there, while his other wives and relatives live in the surrounding ones. By Ndebele tradition when a chief passes on, the oldest son inherits the land and throne."

"Sounds like a typical imperial monarchy."

"If you count wealth in cows and crops." Ian closed the gap between them. "And stipulate the chief's brothers inherit his wives and children."

Tired as she was, she had to grin. "Makes you want to check out the whole family before you tie the knot."

"There are exceptions to tribal customs."

"Such as?"

"You'll see."

"Damn..." She glanced up at Sipho riding on Ian's back. "Darn it, Taljaard, I'm tired of seeing. How about simply telling me for a change?"

"Because a visual stays with you longer."

If it weren't for Sipho's hovering presence, she would have strangled the information from Taljaard's broad neck.

Bold geometric patterns grayed by the starry night accented a plaster wall along the front of the chief's kraal. Ian stopped at an opening and spoke Ndebele to an old man sitting inside. He rose slowly to his feet and stood before Ian, looking first at him, then at the rifle on Joni's shoulder, and last at Sipho.

She figured the rifle would be confiscated, but when the old man spoke again with Ian, the emphasis was apparently on Sipho. The name *Nomathemba* passed between them more than once, and the man gave Sipho a smile of missing teeth and patted him on the back. Sipho seemed the key to more than one puzzle.

Once inside the kraal, they strolled through a collection of thatched structures she supposed belonged to different members of the chief's family. The first two homes were concrete block painted in a similar design as the entrance wall. Farther along, wood and clay huts dotted the edge of a central area.

Tonight, in the midst of the common area, women, children, and men gathered around a campfire and watched six singers delivering a playful song along with dance. They used no instruments except voice, and the pure melody and harmonics sang of perfection. If not for the dire circumstances surrounding their visit, she would have been lulled into complacency.

Children ran up to Ian, chatting in their native tongue. He gently swung Sipho off his back and the kids hustled him over to the fire circle.

Ian circled his shoulders and stretched his muscular arms over his head. He had managed to carry Sipho on his back for miles without complaint. No easy feat even for a solid guy like him.

He slid his hand under the rifle strap at the back of Joni's shoulder. "Sorry you had to haul this, but I don't like being without protection."

His touch had become almost familiar—hand to hand as they had climbed on and off rock ledges, a boost to the backside as they scrambled over and under fences, and body to body as they squished together hiding from prying eyes. Considering she held him at least partly responsible for their messy situation, she should have flinched or recoiled at the contact. That she didn't, raised concern.

To avoid further angst, she slid the pack off her shoulders

before Ian could reach for it. A strap stuck to dried blood and ripped open her cut.

"I'd better have a look at that," he said.

Joni glanced past him, noting a lanky girl staring at them both. Beads dangling in her braided hair reflected firelight. "In a minute. I think we have company."

A skirt with geometric patterns hung above the teen's slender ankles, signifying her arrival to womanhood. She eyed Joni's couture from head to foot. Usually indifferent about her clothes, Joni cringed and brushed at the grime smeared on her. The effort only produced dust.

Ian called out a name, prompting the girl to scoot with childlike enthusiasm to his side. They conversed in a tribal language while she kept a curious eye on Joni. Ian nodded at something, and a pleased smile spread across the girl's face.

The teen motioned for them to follow. Perhaps Ian had asked directions to the washroom. Fat chance. He might be nurturing to Sipho, but she was on her own.

A dirt path led them to the largest dwelling. Inside, Ian hooked his rifle on a wall peg. He bowed politely and then left them alone, taking up a position outside the door.

The house was simple, with no extravagances inside. A single lightbulb hooked on a wall lit the main room. At least this village had power, but so far this was the first sign. The room had a hearth and substantial chair on one side, two narrow tables pushed up against a wall, brightly colored wooden chests, and enough rolled reed mats stored along one wall to seat a few dozen people. Was this the chief's house?

The girl slipped into a back room. She returned holding up a yellow skirt patterned with bold red and orange flowers. Joni understood the gesture. Most native women in the country, particularly the rural areas, wore skirts and covered their legs. Her cargo shorts, although typical of tourists, didn't quite fall into appropriate dress for meeting a chief. Neither did her boots and tattered shirt, but at least the skirt quashed her major fashion faux pas.

With a shy smile, the girl helped Joni wrap the bright material around her waist and tie the ends. Although a bit lumpy over Joni's shorts, it hid her legs and the dirt that adhered to them.

"*Siyabonga kakulu.*" Joni made a stab at thank you, one of the two phrases she had learned on the trek to the kraal.

"*Ziko ndaba.*" The teen smiled, apparently pleased at Joni's attempt to speak her language, before slipping out the door and heading back toward the singing.

Joni stepped into the cool night. Meager light spilled from the doorway and highlighted her ragtag clothes. The skirt was a poor attempt at blending in with the natives, but the teen's willingness to help bode well for their safety.

Ian rested a shoulder against the house, looking relaxed and listening to the voices carried on the night air.

She tugged at the skirt. "You've gracious friends. I'd have preferred purple flowers, but it's all she had in my size."

For the first time away from Sipho, she heard Ian genuinely laugh. He stood and hovered close. "I like that. It shows strength."

"What? The skirt?"

"Your sense of humor."

His compliment made her uneasy with some of the harsh comments she'd made earlier to him. "Look, I...um...need to explain what I said about Sipho earlier. You know, the kid thing. I like kids, I really do."

"Considering how well you and Sipho are getting along, I'd say that's obvious."

"It's not that. You see, I gave up working in situations with kids involved. Lemmon knew that but sent me anyway."

"Because you were the right one for this job."

"But I'm not."

Ian raised questioning brows. "You handle Sipho like a pro. Are you a psychologist? Teacher? Mom?" He paused. "I didn't think to ask if you were married. Just assumed—"

"No. No husband, no kids. Look, accept the fact I was pissed at Lemmon and shouldn't have said what I did. We can leave it at that."

Ian lowered his head close to hers. "I imagine you have a valid reason for your feelings. Considering the nature of your business, I won't ask why. I will, though, ask for one thing."

"Sharing these intimate adventures isn't enough?"

"I want your promise to do what it takes to keep Sipho safe."

His words shoved her thoughts back to a haunting memory. Damp, earthy smells of jungle nights returned as though it were yesterday. She had her arms wrapped around a frightened kid who jumped at every sound and rustle. "Will we make it?" he'd asked continually for the first few days. Her refrain had become, "We'll make it out. I promise." As the days went on, his faith in their survival grew, and he quit asking.

She looked uneasily back at Ian and made the mistake of connecting with his steady gaze. "I don't make promises anymore. I can only try my best."

CHAPTER ELEVEN

Whatever inner turmoil haunted this woman, it etched pain on her face. At one moment self-assured and full of fight in the face of danger, the next she refused to commit to the well-being of a young child. Had he made a mistake going to the Americans or entrusting Sipho to her hands? Before he handed over Sipho, Ian would do whatever it took to ensure he'd made the right decision.

With all the complexities involved in this nightmare, he hadn't foreseen this one, or that the Americans would attempt to soften his concerns and suspicions with a woman—one strong enough in body and spirit to face his enemies…and him. Damn them.

Joni rubbed her arms as though chilled and stepped away. "Sipho's been out of sight for a while. We'd better check on him."

"Sipho's among his people." Ian anchored her in place with a gentle but firm grip on her arm. She tensed at his touch. A surprise since they'd been in much closer quarters throughout the day. "How about I look at your shoulder?"

"It's just a scratch."

"One that will look bad to anyone we meet. There's a proper place to clean it up."

111

"My face and hands could use a wash, too."

He took the opening to touch her again, test her reaction to him without being driven by fear to survive. With thumb and forefinger, he grasped her chin and angled her face into the light. He detected a hint of unease, and felt it himself, all the way to... "Yeah, you could use a little scrubbing before we meet the chief."

Her hand touched his, nudging it from her face. She tried and failed to sound casual. "I take it hoping for a hot bath is a bit optimistic."

"Best to let you check out the facilities."

"Is there a girls' room included?"

"Follow me." With her hand firmly in his, he led her along the edge of the central kraal. Cows lowed close by, and the familiar smell of their fertilizer lingered in the air.

He stopped in front of a long cement-block structure with a thatched roof. Outhouses looked the same in any culture. Indoor plumbing for sewer or water hadn't been extended to individual units of the kraal.

"I'll wait here." He watched her trudge stiffly into the toilets. A glow came from the nearby campfire, as did a rich humming of male voices. Their sound mimicked the sway of the wind through savanna grass and the building crescendo of a thunderstorm.

Minutes passed, which he used to convert his flashlight to a lantern. Why women took so long for such a basic task amazed him. He checked his watch.

"Time's up." He strode toward the outhouse.

Joni stepped out and nearly collided with him. "Looking for me, or making a dash of your own?"

"Simply making sure they had toilet paper, but you're obviously all set." Something about her changing moods had him on edge. She swept confidently past.

They strolled to the opposite end of the kraal, where a water

112

spigot hung on the side of a house. Not exactly a wide-open watering hole. Instead, a worn plastic catch bucket collected precious drips. Remnants of homemade soap bars sat on a rock scooped out to hold them.

"Ah, the master bath." She sighed, likely resigned to a cold-water touch-up.

"The one and only." He dropped his pack and hat, checking in the process no one lurked unseen. "It's where everyone gets water."

"I'm thankful for anything to clean up at this point."

He knelt at the spigot, rinsing dirt off his hands before picking up the soap. He scrubbed and rinsed again, then rose and shook them dry. From the pack's mystic bowels, he dug out a first aid kit and a thin, leathery towel. He stood and tossed the towel over his shoulder.

"That looks like a car chamois."

"Mine dries people, not cars." He pointed to crude hooks on the house wall and slipped the lantern on one. "You might want to hang up your skirt to keep it dry."

A smart woman should have recognized she was about to lose control and, dirty or not, walked away. Instead, with trusting dog-tired motions, Joni untied the skirt and carefully hooked it out of the way. "Why is it men are always trying to get a woman's clothes off?"

"Don't know." Her teasing question sparked guilt. He shoved it away. "It's always been the other way around for me."

A side glance indicated she hadn't completely bought his innocent desire to help. Perhaps she was smarter than he thought.

"Is this the best place?" she asked, attempting to postpone his medical attention. "There's not much light here."

"There's enough." Something in his voice dared her to hold her ground and risk his touch. He tapped a button on her shirt. "I have to reach the wound to clean it. I'll let you do the honors."

She fingered the button. She undid one, then another. On the third, he couldn't help but smile.

She turned away so her damaged shoulder faced him and then lowered her shirt down her back and shoulders.

His fingertips brushed against her skin. Goose bumps rose along her arms and back.

She shifted uncomfortably. "Didn't realize the night air was so cool."

He gently lifted her bra strap near the cut and slid it down off her shoulder. Going through his mind were fifty ways to warm her up.

With dampened gauze, he dabbed around the wound, cleaning away dried blood and dirt. "You were right." It took effort to sound as though her standing half-naked before him had no effect. "It's barely a scratch."

"If you ever talk to Mr. Lemmon, play it up a bit. I'm asking for a combat bonus after this adventure." She winced as he applied a stinging antiseptic wipe and then slathered ointment across the cut. He used little butterfly strips across it before taping on a bandage. "Should heal up right quick. Until then, you'd better keep it clean."

"Thanks, Doc." She started to tug her shirt back up.

"Not done yet." He tore a piece of duct tape and stuck his hand up the back of her shirt. She strained to look over her shoulder. He fiddled with the material before sticking the tape to the inside.

"Done." He carefully lifted her shirt back in place. "Good as new."

"Men and their duct tape." She faced him and fumbled with a button.

"Need help?"

"We've a mission to accomplish, Taljaard. You need to focus."

"I am."

Too bad there wasn't better light to read her eyes. Leaving her off balance would play to his advantage. She couldn't be sure if he teased her to lighten the tenuous circumstances or whether he was being serious. Hell, he wasn't so sure himself, and that was damn dangerous. Kagona had known a pilot was with him, a fact only a few people had access to…and she was one.

Joni finished buttoning, then reached back to check his work. She appeared satisfied. "You realize we wouldn't be in this predicament if things had gone as originally planned."

Ian crossed his arms and stepped forward, forcing her back against the cool painted concrete block of the house. "I don't know whose plan you are referring to, but mine was never to have Sipho at the plane. I'd have been foolish to risk him on the chance your little craft attracted unwanted attention. Your CIA knew that. They're even the ones who reminded me this was a trial run for their experimental plane. They told me not to worry, you were well acquainted with fieldwork."

"Lemmon exaggerated. I've only been in the jungles of Colombia. Not the wilds of Zimbabwe."

"Don't worry, that's my expertise. I'll get you and Sipho back to Taz in one piece." He ran two fingers down her arm and then brushed the tips together as though cleaning off dirt. "You've work to do before I introduce you to the chief. Take your boots off and let's get those arms and legs clean."

She untied her boots and slipped off her dusty socks, stuffing them into a boot to keep dry. She stepped on a square of cement placed in front of the spigot and knelt. Ian adjusted the spigot until the slow drops became a steady but light stream. She let the water run over her hands and up to her elbows.

"Sipho thought I was a bit white to play hide-and-seek on the savanna."

"He was right."

Ian dropped a piece of soap on her palm. She rubbed her hands about it and spread the lather from her wrists to her elbow.

Dirt mixed with water and dripped off her fingers and skin. He knelt and cupped water in his large hands. He dumped it over her soapy skin while she scrubbed. In short order, both arms from the elbows down were cleaned.

"Is this spa treatment tribal custom?" she asked.

Ian coaxed her to a stand and dried her arms with the towel. "Actually, it's a way of saying thanks." To her surprise, he knelt and tossed water down her leg. He began lathering his hands again.

"I can do that." She stretched down for the soap. The bandage on her shoulder pulled, stopping her motion.

He caught her grimace. "That's why I'm on cleanup duty. That cut could have used a few stitches."

She drew in a long breath as his hands slid up to her thigh, carrying water and soap, which dribbled back down her leg. A new set of goose bumps arose.

Soap made his hands slick against her skin. "You didn't have to take the chances you did today," he said.

"I didn't have any choice but to go along, did I?" She bit her lower lip when his thumb trailed down the edge of her calf muscle.

"You had a choice to save yourself and didn't." He lifted her foot and rubbed soap between her toes.

Giggles burst forth. "Damn it, Taljaard, I'm ticklish."

He set her foot down and proceeded with the rinse cycle. "You wanted to know more about Sipho."

"Why is he so important to Kagona? All this can't be over bank accounts and blood diamonds."

"It's that and more. Sipho's father knows where government officials and bankers have squirreled away millions of dollars meant for the Zimbabwean people."

"News flash—party leaders haven't exactly hid their extravagance from the people. It's common knowledge."

He started on the other leg. "It's not Africans the politicians are afraid of, it's the Americans. They take a personal approach to blatant greed at the expense of human rights. They sanction leaders via their pocketbooks. Knowledge of account numbers and asset locations is a tactical advantage."

"Sipho's father was the finance minister. Don't you think it's likely he skimmed funds, too? And if he did, what made him change his mind?"

"I'd like to think his conscience suffered. But truthfully, I believe he did it for his wife. Those suffering most are from the tribe not in power, the Ndebele."

"Nomathemba's tribe?"

"Yep." He shoved away nightmare visions of how his beloved friend had died. Someday, he'd have the final word with Kagona. "The politicians' theft was so blatant, Mukono easily followed the money trail. Eventually he compiled a long list of accounts and assets."

"So how does Sipho fit in? Does he know where this info is located?"

"More or less." He'd share the *more* later when assured of her alliances.

"Why did his father tell you about the list? I know you and Nomathemba were friends, but you're white and part of the minority from which his people had wrested political control."

"I had connections he used to trace the money trails."

"How did Kagona find out?"

"Mukono spooked party leaders when he spoke against skimming tax money from recently discovered diamond fields. Kagona tracked his actions and figured out what he was doing. Kagona's cronies plotted to snatch Mukono and his family so he could learn who Mukono had told and what data he'd given them. Kagona did a sloppy job."

The glide of his hand while he flushed water across her skin left him imagining the uncharted territory hidden beneath her shorts. He looked up to find her eyes focused on a distant star.

"Where did you learn to manipulate people so well?" she whispered.

He chuckled. "That answer is for the next spa treatment. Quit talking and relax. I don't do this for just anyone."

He picked a tight place on her calf and massaged it.

A soft moan passed her lips, the kind saying he'd provided relief. "That feels good. My leg was getting stiff on the walk here."

"I saw you favoring it. Now I see why. I'm working around a bruise I found underneath all the dirt. Was that from the tussle with the helo?"

"At least I'm still alive. Have you ever considered a safer lifestyle?"

He ignored her comment. "So what do you think of Sipho?"

"Smart. Seems awfully fond of you."

Ian smiled. "Just like I am of him. While I don't agree with all his father's political views, Mukono is better than most in this government. He also is a good father to Sipho and loved his wife. I promised Sipho's parents I'd take care of him, and I will."

"Promises are a risky proposition. What if something happens to him?"

"It won't." He scooped handfuls of rinse water onto her skin.

"Brr. How did Sipho get access to the information? And why did his parents even—"

He slid his hand under her shorts to the edge of her panties. "Sorry. Slipped."

She stepped away. "You did that on purpose."

He made no effort to hide a guilty smile.

"Move out of the way and let me wash my face."

He lifted the bucket so she could scoop water without bending. Once done, he set the bucket down, and she reached for his towel.

He held onto it. "There's a trick to using this."

With slow precision, he wiped the towel across her forehead at the hairline, forcing her eyes closed as water dripped down over them. He skimmed along her brow and dipped into the crevices near her eyes. He was close, much too close, and wanted to get closer.

She reached for his hand. "Ian, I don't think—"

"I'm almost done." He wiped across her upper lip to the edges of her mouth. He let the cloth linger there…tempted to do more.

Drawing in a slow breath, he stroked under the edge of her ear and moved slowly toward her chin. Her body shuddered. The singing from the central kraal grew louder and faster…encouraging. The cool night grew warmer.

The thumb of his hand holding her chin brushed against her lips. "Now's the time to say no," he whispered and reached an arm around her waist.

Her eyes flickered open and caught him inches away looking for an answer. If she had one, it stuck in her throat. She slipped her hand to his neck.

The kiss grew from a tender trace of his tongue over her teeth to demanding and seeking. He wanted more. A lot more. Warnings to slow arose from within. What if this was part of her agenda?

The singing stopped, and the real world intruded. She gently pushed away. "I don't think this is part of the mission plan."

"Then maybe you didn't get fully briefed." Ian grinned and made a visible effort to compose himself, but his breathing was slow to cooperate. "Perhaps at a better time, I can give you another update."

She wrapped the skirt about her legs as if that would ward him off. "What I really need is to learn everything I can about Sipho. Any mistake I make not knowing the situation could cost us our lives."

"No problem," Ian said. "How about a short stroll tonight after Sipho falls asleep?"

"What about now?"

"The chief is waiting to meet us."

"Later then, but I want answers, Ian, not playtime."

"Promise."

Joni grabbed her boots and found a dry spot to sit. Ian watched, drawn to analyzing everything about her. Damn it to hell, his psyche was a mess. With a shake of his head, he packed up the medical stuff and slung the pack casually over one shoulder.

"Come on." He collected the little lantern. "Time to get an official invite to stay the night."

CHAPTER TWELVE

Exhaustion racked Joni's body as she trailed after Ian. She needed downtime and soon. In one very long day, her physical and emotional states had been stretched to their extremes. She longed to trust this man, but the last few minutes had displayed her vulnerabilities around him. Was this part of Lemmon's scheme?

"So are we groveling for protection from some old chief with a half dozen wives?" She rubbed her arms as they walked toward the roaring bonfire. Its flames promised warmth.

"The chief is young and single. Oh, and I need a name."

"A name for what?"

"You. What's your last name?"

"We kissed and you don't even know my name?"

"Is there a rule against that?"

"Bell. Joni Bell."

He gave her a two-dimple grin. "Okay, Joni Bell, stay here until I signal you to join me."

Ian walked toward the assembled tribe. He showed signs of wear, but also appeared relaxed. He seemed to have great faith in these people.

Not in a mood for introductions and pleasantries, she waited only a minute before plodding after Ian. The tribal leader, still dressed in the ceremonial leopard cloak, white pith helmet, and black staff, wasn't hard to pick out in the group ahead. With his back turned toward them, he appeared rather short, but then heredity, not physical prowess, picked number one.

Ian said something, and the chief whirled around. Joni stopped in her tracks. The new chief must be one of those *exceptions* Ian had mentioned. Heck, the attractive woman looked to be in her midtwenties.

For a brief moment, the chief appeared ready to throw her arms around Ian's neck. Instead, she presented him with a genuine smile of both surprise and pleasure. The chief stepped closer to Ian and they spoke. He indicated Sipho among the kids. The chief bowed her head, and Joni could have sworn she wiped away tears. Ian continued to speak to her, and once she regained composure, he pointed in Joni's direction.

They looked toward her, and Ian signaled Joni to approach. The chief eyed her with a bit of skepticism until Ian whispered something in her ear. A frown crossed the chief's face, but she nodded at Joni and appeared relieved at her presence.

Joni reached Ian's side and bowed politely to the petite leader, since Ian had forgotten to mention any specific protocol.

"Chief Zwide," Ian introduced, "this is the woman seeing to Sipho's safety. Miss Joni Bell."

"You are brave woman, Miss Bell," she said in decent but imperfect English, "for helping in such difficult time. Nomathemba was of our tribe and family. A good woman, like you. She would give you many blessings for saving Sipho. She encouraged me for getting schooling. In the city, I teach others like me. Now I also lead my village. Any who help one of us is welcome here."

"I'm sure the people are proud to have you as their leader. I will do my best with Sipho."

An older gentleman interrupted and spoke quietly with the

chief. She nodded and turned back toward them. "I am sorry, but I must go. Sibongile, my sister, will have place for you to rest. Please enjoy the *pungwe*."

Joni stood next to Ian until the chief moved out of earshot. "I like to see women making inroads into society."

Ian thoughtfully crossed his arms. "Not everyone in the tribe is happy about her ascension. Her father had no sons, and his only living brother lacked chief qualities."

"The black sheep?"

"Several visits to the local tank. The chief has to settle local and tribal disputes. A criminal history disqualifies a candidate. Her father nominated her, and the tribe more or less agreed."

"Not all agreed." A deep voice came from behind them. A man stepped up to Ian's side. "A woman have no right to chief. A chief must be warrior. A woman cannot have *intelezi* and know ways we fight."

Joni wanted to ask why in the hell a woman couldn't have whatever *intelezi* was, but Ian struck up a conversation and purposefully blocked her out. He made it obvious he didn't want her speaking to this man, perhaps because he might recognize an American accent. Not a bad idea, as their presence among a crowd seemed a risky proposition at best. The man chatted on and apparently saw nothing rude in Ian's exclusion of her. She hated to think it might be normal treatment for women here. It hardly mattered. She had no desire to discuss anything except where to put her head down for a good snooze.

She left the men and settled on the ground among a crowd of at least thirty people watching the singers. Their naked chests, animal-skin groin cloths, furry shin guards, and plumed headgear hadn't changed much from their Zulu ancestors. Hand them a long shield and spear and they'd look like perfect warriors. They not only sang but leaped, stomped, shimmied, and kicked. Primitive, invigorating, and sensual all in one. As if cued by her thoughts, Ian slid in close to her.

"I see not everyone in the family is happy with a lady chief."
The heat from his body warmed her side. "Rather makes our visit
riskier, doesn't it?"

"Local squabbles won't endanger us."

"You're quite confident of yourself. Ever been wrong?"

"More often than you know."

"What is this *intelezi* the guy back there mentioned?"

"It's war medicine sprinkled on soldiers. A *sangoma* cooks
up the special herb brew. Women are not allowed to learn how it
is made or have it touch them."

"What or who is the *sangoma*?"

Ian pointed at a figure directing the singing group. His close-
cropped, graying black hair labeled him as one of the elders.
"He's the village's herbal doctor."

Joni had gone from flying the latest aircraft to ancient
African traditions all in one day. Any time now she expected H.
G. Wells to step out of his time machine.

"I had *presumed* witch doctors disappeared with Dr.
Livingstone. I realize people have superstitions, but surely the
entire tribe doesn't believe that stuff."

"Are all Americans so sarcastic?"

"Probably. But it only shows when they're dog tired. I didn't
sign up for anything but a plane ride. Someone forgot to brief me
about a foot tour of Zimbabwe."

"Don't underestimate the power of ancient beliefs.
Regardless of this being the twenty-first century, superstitions
are alive and well in Africa. Tribesmen regularly hunt and
destroy witch charms hung in the forests. Some even believe the
last drought was caused by Christians plowing on a tribal holy
day."

The song ended and the singers gathered close to their
director, the *sangoma* Ian had pointed out. Dressed in brown
pants and a loose, colorful, patterned shirt, he didn't look much
like a witch doctor. He spoke in a tribal dialect.

One man began to sing again. Bit by bit the others added their voices to an intricate a cappella harmony. Some parts reminded her of music from a children's movie. Now she knew where American songwriters had come for inspiration. "What are they singing about?"

"Never heard this one before. Most songs are ballads about animals, war, or tribal life. It's a style called *imbube*. The *sangoma* announced these men had worked for a month learning the patterns he created for this finale."

A beautiful harmony wound its way into her soul. Sometimes repetitive and other times complex, the clean, clear voices produced their own symphony. When they finished, the audience was silent for a long moment until cheers began.

The children hopped up and surrounded the singers. The tallest performer swept up Sipho and let him perch in his arms. Sipho tapped him on the shoulder and then started to talk.

The smile on the performer's face changed as his jaw dropped open. Sipho kept right on moving his mouth in what seemed a one-sided conversation. The performer waved at those around him as though hushing them. More and more people stopped making noise and turned to listen to Sipho.

Whether good or bad, whatever was happening brought attention to them. That couldn't be good. Joni grabbed Ian by the arm and pointed to Sipho. Ian must have agreed with her analysis as he sprang to his feet.

A clear, childlike voice reached her ears. Sipho sang, in perfect pitch, one of the complex patterns from the finale. Most people exhibited wonder and smiles, but the *sangoma* had sheer horror on his face. It probably wasn't a smart thing to piss off a witch doctor. All they needed were a few hexes to add to their problems.

Ian tapped Sipho on the shoulder, ending his singing spree before lifting him from the performer's arms. Sipho waved to cheers as Ian carried him away. Joni didn't like the concern written in the creases on Ian's forehead. Had his confidence in their safety here waned?

Quite content and pleased with his performance, Sipho slid from Ian's hold and gave her a hug. She crouched to see him eye to eye. "One of these days, you're going to have to teach me how you did that."

"It easy. Mama makes me sing. Say song make heart happy."

His innocent words dug deep into her emotional territory. "Your mother is a wise woman."

"She best in whole world." The happiness that had brightened his eyes and rounded his smooth brown cheeks dissipated. His gaze dropped to the ground. "Ian say bad accident hurt Mama. Angels take her. *Sangoma* say no angels. She with spirits. What you think, Miss Joni?"

"I am not as old or wise as the *sangoma*, but I know angels like to sing. I think they and your mama were listening tonight."

While her heart ached, the edges of Sipho's mouth rose in a contented smile. He gave her another hug before leaning off balance against Ian's side, one scrawny leg twisted over the other. Ian affectionately buffed the top of Sipho's head.

"We're staying with Sibongile tonight," he said to Sipho. "Can you find her for us?"

Sipho skipped off into the gathering. Ian carefully observed the group, the *sangoma* in particular, as he spoke with members of his chorus.

"You looked worried."

"The man who was holding Sipho is from a nearby kraal."

"Trustworthy?"

"I plan to sleep with one eye open tonight."

"Great. Should we leave?"

"I won't be caught unaware. I have an exit strategy."

She rubbed her arms, wishing like heck for a safe hole to crawl into and sleep. The mush she had for brains right now had no ability to discern the wise moves to make.

The chief's sister, whom Ian introduced as Sibongile, and her husband escorted the three of them to her home. The round thatch-roofed hut with one room was a simple home with one bed, a sideboard, and a wooden trunk. Outside the door a cooking pot hung on a tripod.

"Newlyweds." Ian pointed at the freshly painted blue, green, and yellow motif on the curved inside wall.

"Welcome." Sibongile hesitated as though wishing to say more but not confident of the words. Instead, she smiled and pointed toward a large mat on the ground and mimicked eating. Guess a dining table came with future anniversaries. Ian leaned close and whispered that she was going to prepare food for them. Joni's stomach growled thanks.

The husband chatted in Ndebele with Ian, pointing in various directions, and, if she had to guess, talked farming and politics. Sibongile stretched a hand hesitantly toward Joni's head and then drew back. Joni smiled and nodded. Sibongile reached up again and plucked out a stick.

Okay, so the bad hair job scared the locals. Guess she should take that up with her spa manager. Good thing no mirror hung on the wall to reveal how tattered she must look. Sibongile mimicked washing her hair, and Joni nodded agreeably. Sleeping on a dust bowl hairdo wasn't good for the lungs.

Sibongile collected a large wooden bucket and led Joni back to the water spigot. The fire had almost died out, and the chief's friends and family were retiring for what was left of the night. Far off, nocturnal bird and animal calls floated in on a light breeze. Sibongile filled the bucket and placed it on a pile of rocks set up for such a need.

The chief's sister pointed out soap bars before motioning something about food and bowing out. At this point, any munchie, even mopane worms, sounded appetizing.

Joni snatched up a water scooper, then paused. Voices mumbled in the distance. Soft feet padded quickly and with purpose somewhere in the shadowy kraal. Joni had no way of knowing what was typical and what was not. She had to quit being on edge. Ian was with Sipho, and he trusted these people.

Still, she looked into the night all around her before grabbing the bar of soap, flipping her head upside down, and scooping water over her hair. From nowhere, the dark cloth of a skirt swooshed against her side. She jerked up in surprise, but firm hands held her head down just above the bucket.

"My sister told me you were here," the chief whispered as fingers snatched the soap. "Keep your pale arms down. Best to hide them." She started kneading soap into Joni's wet hair.

"Chief Zwide." A voice called out, not deep enough for a man, but older than a mere boy.

The chief kept her back to the person and shielded Joni's white face and limbs. "I not hear you request to enter my kraal, young Jabu." She spoke in English as though to reiterate her stature as a teacher as well as chief. "But you are welcome without weapon."

Shit, some strange kid had waltzed into the kraal with a gun and she had her head in a bucket. At least they were in shadow. She tried to lift her head, but the chief kept firm downward pressure.

"Some, ah, say—" he attempted in English, but broke into Ndebele. All Joni recognized was the word *imbube*. Had he come to hear the singers?

The chief answered in kind, obviously telling him he'd missed the show, and by the feel of hand gesturing going on, complaining he was disturbing the women engaged in domestic necessities. Joni's back ached. She needed to stand or at least rest her hands on the bucket.

"*Uhambe kuhle*, Chief Zwide."

The chief repeated the youth's salutation and the footsteps walked away. After a few seconds, the chief whispered, "You can stand."

She wrung out her hair and stretched tall. Stiffness but pure relief came with it. Joni kneaded her back muscles. "Who was he?"

"From nearby kraal. He belongs to local youth brigade."

"You're afraid he might reveal we're here?"

"Brigade is bad for tribe young people. Government takes them away to train. Dresses them in uniforms. Mothers and fathers are not allowed to see sons and daughters. Many boys go bad. Rob and beat in name of government. Girls—" The chief paused and sighed. "I cannot tell you shame. I sometime see brigade at roadblocks on way to city. Jabu was good boy before, but I do not trust now."

"Ian and Sipho?"

"They are safe. Finish with hair and we will go."

Joni rubbed more soap into her hair and scrubbed out some of the dirt. She poured water over her head and then squeezed it out. Unease had set in, and she doubted any rest would come tonight.

When she straightened, the chief handed her a large-toothed comb. "Come. We walk."

Joni strolled with the chief back toward Sibongile's hut while combing wet tangles out of her hair. The kraal, flavored with burned wood and lit now only by celestial light, had grown relatively silent.

"Ian says you will take Sipho to South Africa until father free. Sipho will be lonely without Nomathemba...without family." The woman who had lost her friend, not the chief, was speaking. "Can you keep him safe there?"

"I'll do my best." She hesitated. The chief looked at her for more reassurance. "I promise."

Hoping the promise wasn't a hollow one, Joni handed back the comb. A cool night breeze had practically dried her hair. She stifled a yawn.

"Sibongile is waiting to feed you and I must let you sleep. I will cherish your promise to Sipho. Bring him back when safe.

Times are bad now, but I hope better soon. *Litshone njani.*" The chief held out her hand.

Joni shook it, praying she could live up to this believer's expectations and surprised that it mattered. "Good night."

She entered the hut and let her eyes adjust to a brighter setting. A single light of low wattage hung over the cooking tripod that had been moved inside.

Ian sat cross-legged on a large reed mat with Sipho next to him. Joni wadded her skirt up around her waist and joined the men. She probably had broken the appropriate custom, whatever it was, but hoped as a visitor she'd be forgiven.

"You guys okay?"

"There were plenty of places to hide if the boy had wanted to search," Ian said. "I don't think that was his intention. The brigade members are young and travel in groups to keep an eye on each other. Make sure no one runs home. He came alone. A risky proposition for him."

"You sure?"

Ian raised his brows and stared hard at her. "I'm sure."

Sibongile served them corn porridge Ian called *sadza*, adding apologies for the lack of meat to make it a stew. Sipho gobbled his down, and Joni was grateful for anything to fill her stomach.

A young boy who had been playing with Sipho earlier appeared carting a bucket of liquid. He called it *utshwala* and said it came from Chief Zwide.

Sibongile poured some into two chipped enameled mugs and handed them to her guests. Ian raised his mug but didn't drink. Instead he looked expectantly at Joni and waited as she brought the brew to her lips. The hazy light made the dark liquid appear chocolate brown.

"Go ahead. You'll like it," he said.

Suspicious, she sniffed and swirled it, noting it had the consistency of thin gruel. "What is it?" She maintained a smile so as not to offend their hosts.

"Beer."

She sipped at the liquid. Not top-of-the-line brew, but refreshing. Once they were satisfied, their hosts partook. As food and drink filled Joni, her eyes wanted nothing more than to shut. She made no complaint when the hostess pointed out three rolled woven mats. Two leaned against one part of the curving wall, and a third was a good distance away. Guess the unmarried boys and girls didn't sleep together. Not a bad idea in this case. The last thing she needed was a mass of testosterone hovering inches away from her body.

Ian put his lips near her ear. "I'll get Sipho to sleep, then we can walk and talk."

He stretched out on a mat with his face hidden under his hat. Sipho curled against him, content with the security of his presence. Sipho's faith in Ian wasn't misplaced. Joni had no doubt he would and could kill to keep him safe.

She shook out her mat and placed it so the boys were in sight. She rolled up her borrowed skirt to tuck under her neck. Definitely not first-class accommodations, but it was a place to lay her head. How safe a place, time would tell, but she planned to sleep lightly.

Darkness engulfed the room as the chief's sister extinguished the single light. The bed rustled as she settled in with her husband. The newlyweds seemed used to the lack of privacy, probably a consequence of living in a family hamlet.

In no time, the regular in and out of Ian's breathing carried across the room. His performance to encourage Sipho into a long conversation with the sandman was effective. Her own eyes sought to remain shut.

Sipho let off a long sigh and drifted off to sleep. In another minute, she'd nudge Ian and sneak outside for the awaited rendezvous and briefing. At last she would find out exactly what Sipho had locked up in his head and why.

A cool breeze wafted in from an open window and the door.

SANDY PARKS

Her boots weighed heavy on her feet. She had questions for Ian. They needed to talk. What were the questions? Her brain refused to focus.

The day's scenes filtered through her thoughts—the gun clicking empty in her face, Ian carrying Sipho, African singers, and a kiss. And what about Taz? She had to find Taz.

Fatigue glued her arms to the mat. She had to get up. But she didn't want to move. Or couldn't? Fear swirled then drifted off. If she fell asleep, Ian would wake her.

CHAPTER THIRTEEN

B right light filtered through Joni's eyelids. Laughter and voices echoed into the room from the outside. She cranked her unwilling eyes open and recalled how she had arrived at this hut.

She sat up, perhaps a bit too quickly, causing her head to throb. Last night's beer packed an unexpected punch. She'd slept like a rock with no memory of waking during the night.

Ian still stretched out on the nearby mat. She'd expected Mr. Wildlife to be up with the sun. From the full sunshine visible out the door, the morning had started hours ago. Sipho must have sneaked out earlier and joined the children playing outside.

So much for the promised answers about Sipho and his parents. She rose and stretched, making no attempt to keep quiet. Ian didn't stir. She nudged him in the side with her boot. He jerked up, catching his hat as it fell off his face.

Quickly he looked around.

"Where's Sipho?" He snatched up the school pack as though the boy hid behind it.

"Out playing with the other kids, I assume."

"You assume?" He dropped the pack and sprang to his feet.

She recoiled at his tone that implied she was a toddler's

appointed babysitter and had left the door open to a lion's cage. He scrambled out the door before she could utter a word in her defense.

She tore after him as he checked the children playing nearby. He stuck his head into one structure after another in the kraal without pause. "Weren't you the one who said this place was safe?"

"How's your head feel?"

"Like I slept on a rock last night. Why?"

He circled the toilet block. "I haven't slept past sunrise in the last two years. Sipho doesn't leave my side without telling me first. That includes waking me up when he needs to pee."

A sick worry nibbled at her conscience. Horrid images resurrected of the Colombian boy within her sight, fishing at the river for their dinner. She couldn't save him. The nightmare couldn't be happening again. Wasn't once in a lifetime enough?

"Sipho," Ian called with hands cupped beside his mouth. Those in the kraal turned to stare.

Ian searched their questioning faces, but no one showed any hint of having seen the boy. They caught Sibongile leaving the toilet blockhouse and questioned her about Sipho. Joni had to wait until Ian translated.

"He was gone when she rose this morning. She assumed he'd gone to play with the boy who had brought the beer last night." For such a tough guy, distress washed across Ian's face.

"Don't panic, he's likely playing with the kid now."

Ian tore off.

She caught up in time to see the boy's mother shaking her head. Joni's hope came crashing down. "What did she say?"

"Her son has left for school and won't return for hours. He'd been home all morning. Sipho was never with him."

"He has to be here somewhere. Where do they go when not in school?" Her wavering voice reflected the cancerous doubt eating up her insides.

"Sipho wouldn't wander off. He knows what happened to his parents and how Kagona's men invaded his house. Kids, even like Sipho, don't forget. He knows I'm his protection." He focused on the chief's house.

Tempted to remind him it had been his decision to trust these people, she held her tongue. Placing blame wouldn't get Sipho back if he really had disappeared.

Ian aimed toward the chief's place, while Joni swept a wide circle of the kraal. Nothing but normal village life. No guys with guns or hovering choppers or struggling kids. By the time she rejoined Ian, he spoke with someone at the chief's house.

The two men exchanged lengthy words accompanied by arm and hand gyrations. Finally, the man shrugged and disappeared. Ian didn't move.

"What did he say?" she asked.

"The chief is meeting with village elders. She can't see us, but with luck the man will get a few answers."

Ian paced each anxious minute until the man returned. Ian's face grew taut as the man spoke. Once again the exchange became lengthy. Ian suddenly thanked the man and took off across the kraal.

Tired of dogging behind him, Joni jogged at his side. "How about a little translation here. I left my phrase book back in the hotel room."

"The chief didn't send us beer last night and only one person here had it available. His hut is straight ahead."

Ian barged into a large thatched hut. Tables covered in jars and pottery filled the large room. Strings of dried herbs and plants hung from the ceiling. An old but functioning woodstove heated an array of steaming pots sitting on top. The heady mix of pungent scents was nearly overwhelming. Ian stopped a pace away from the *sangoma*.

Hunched over a table grinding ingredients in a mortar, the herb doctor straightened at their intrusion. "You no welcome. Go."

Ian kept his hands at his side, but his fingers opened and shut into a fist. "Where is Sipho?"

Sangoma shook his head. "Wicked boy."

"You sent us the beer last night laced with a sleeping concoction. I'll ask you one more time, where is he?"

Rage filled Joni as she pushed between the men and into the doctor's face. "*You* spiked our beer?" Her heart turned to stone as familiar feelings of helpless loss emerged. "We're talking about a kid here. What kind of man are you to harm an innocent child?"

"Boy not right. Evil follow him."

"The only evil—"

Ian lifted her out of the way before securing a handful of the *sangoma*'s loose shirt.

A commotion behind them brought the chief into the hut. Without a word, she signaled Joni and Ian outside. Even though she was a young woman, her position held weight. Ian released the man and stepped away. He prodded Joni out the door.

Inside, a conversation started in a low tone and quickly rose in volume and speed of delivery. Tongues clicked and hands gestured.

Ian pressed against Joni's shoulder as they both squeezed in the doorway watching the scene. The chief's face grew graver with every word. The news didn't look good. Ian grew tenser and more enraged with each phrase the witch doctor uttered. If only she could understand. The conversation quieted. After several minutes the chief stepped outside and motioned them away from the hut.

Joni's stomach knotted. They had to find Sipho. He was her mission. He'd saved her life back on the reserve and, damn it, she cared for the kid.

Chief Zwide looked unhappy with the circumstances. For a second, she avoided looking straight at them. Then, as though remembering her newly bestowed position, she took a deep breath and lifted her chin.

"My apologies for *sangoma*'s actions. He believed he was protecting kraal. He said only person cursed by witch could repeat long and delicate melody after hearing once. He feared such thing on night of chief's coronation would bring bad luck to village."

Joni tapped her fingers against her arm. They needed details—fast. "There's no time to worry about our mistakes. Where's Sipho?"

The chief's jaw started to quiver. "Ian, I failed you and Nomathemba. I should have seen threat to Sipho." Her voice faltered and she looked up at Ian. "It is my fault child is gone." Moisture glazed her eyes.

Ian wrapped a brotherly arm around the chief's shoulder. "The fault is mine, not yours. He's my responsibility. I'm the one who relaxed my guard."

Chief Zwide appeared heartened by the belief the fault did not lie entirely on her shoulders. "Nomathemba was my friend. She is one who made me strong." She straightened from Ian's comfort. "The *sangoma* thinks Nomathemba was wrong to marry into Shona tribe. He saw Sipho as Shona, not Ndebele. I knew that was true."

"I heard most of what the *sangoma* said, but not who has Sipho."

The young chief glanced over her shoulder toward the door, obviously ashamed of her tribesman's action. "Trader came to him. Claimed boy in kraal. Told *sangoma* child was big trouble for us."

"The trader Brugman?"

She nodded. "He takes problem things out of my people's hands. Told *sangoma* if boy didn't go with him, then police or men with guns would come soon."

"The *sangoma* could have come to me. We'd have left."

"I am chief. He must talk with me. But he did not. He believes my friendship with Sipho's mother will block my wisdom. He thought men with guns would be angry if you took Sipho from kraal. He worried what they might do. Thought it was best to give evil to trader."

"Sipho's a kid, not an object," Joni interjected. Sipho's innocent smile flashed in her head. He was a pawn in deadly game. "We could've protected him." Hope fled. Two years to bury the memories of the Colombia incident, and now the same terror was happening again.

"*Sangoma* was afraid Ian would fight, so put all in hut to sleep."

Ian shook his head, looking as disgusted as she felt. "Brugman has a ruthless reputation. Your tribe would be wise to do no further business with him. He and his men are more evil than all the witches of Zimbabwe put together."

The chief took Ian at his word and looked even grimmer at his assessment. "Sipho is not evil and Nomathemba is in my heart, but I understand beliefs and weaknesses of my people. Even if wrong, *sangoma* does these things for our good. Please, help fix my mistake. Find Sipho."

Ian's blue eyes grayed with darkness of purpose as though he would fight to get Sipho back or die trying. "We'll need to get to Bulawayo."

"I have ended my meeting. I can drive you to city, but we must be careful. More roadblocks with election coming. They know day is too late for me to go teach. They may look hard at car." The risk was as great for the chief and her village as it was for them.

"We don't have a choice," Ian said. "We have to move fast."

Events were going poorly on the chief's first day, and it played across her face. She had alienated the village herb doctor and lost one of her tribe. Joni found it hard to sympathize. *Welcome to leadership.*

"Bulawayo we go, then. I will need ten minutes to prepare." With a somber nod, the chief left.

From the distant doorway, the defiant *sangoma* cast Joni an evil eye. Last night's spiked beer made her head pulse. She narrowed her eyes and fired him one back.

"How could you be afraid of a child?" she called out.

Ian stepped in the way, turning his back on the *sangoma*. "Let's go, Joni. We have bigger problems."

Such callousness with a child's life defied imagination. She strode toward the closest home in the kraal and once out of the *sangoma*'s sight slammed a fist against a rough wall. She wasn't losing another kid, not one whose innocent eyes had already seen enough pain and suffering. She leaned against the wall, fighting the implications bombarding her brain.

Ian lifted her scraped knuckle. "Feel better?"

"No." She pulled her hand back. "I want to blame you and your tribal friends for his disappearance." Tension had run her mouth dry. "But I can't. I came along with you. I was too exhausted to argue."

"Joni, this mistake was mine. I failed to cover the contingency of me being incapacitated."

She examined her hands, and shook her head. "Nothing makes sense. You heard what the chief said. The trader came to the witch doctor. Not the other way around. What do you think that means?"

"Someone revealed our presence."

"My guess, too. Is Brugman really dangerous or did you say that stuff to scare the chief?"

"He's a rich bastard who exploits African nations in crisis." Ian scanned the kraal for threats for the second time in a minute. "Zimbabwe is his latest target. He plays any side of the fence that's profitable."

"Convenient for him."

"His victims believe he is there to assist. The more desperate a country becomes, the richer he gets. He feeds on misery and will create it if necessary. On a positive note, he's not fond of killing unless it adds to his bottom line."

She rubbed a temple, pressing away images of the Colombian boy. Why were such innocents used as pawns? "I hope that comment wasn't supposed to ease my anxiety."

"It wasn't."

"Is Brugman friends with Kagona?"

"Friends, no. But Brugman will work with whoever pays him."

"So you think Kagona hired him to find us?"

"I doubt it. Kagona had us in his sights. With us that close, he wouldn't want to pay Brugman's steep price to get Sipho from him."

"So how did Brugman hear about the hunt for Sipho?"

"This was Nomathemba's home. Kagona likely put the word out to local police units in this area to watch for us. It's typical of Brugman's operations to have inside contacts. That's exactly the type thing they'd pass along."

"So who sold us out to the police? The *sangoma*? The singer from the other kraal?"

Ian worked a fist against his palm. "I haven't figured that out."

The pieces of this jigsaw puzzle had doubled in number, and none fit together well in Joni's head. "How did Brugman beat out Kagona to snatch Sipho?"

"I injured Kagona during a scuffle at the farmhouse. He would have required medical attention. Our capture is personal. He'd want to be in on the action. I'm assuming that delay gave Brugman the opportunity to slip in."

Joni struggled with the facts. "The timing was tight from our arrival to the time the boy brought the spiked beer. Regardless of what the *sangoma* claims, he must have called them."

"Except there are no landlines around here and cell reception is nonexistent."

"Two-way radio?" Joni slipped a hand into her pocket. Her fingers encountered cool metal.

"Possibly. I can ask Sibongile if the kraal has one." Ian shifted closer; his features had gone stone cold. "You haven't asked about satellite phones."

"Okay. Anyone in the tribe have one?"

"I doubt it." He touched her shorts over the pocket where her hand resided. "But you do."

"Don't be ridiculous." She withdrew her hand and swiped his away. "I'd never turn Sipho over to those killers, and this phone hasn't been out of my sight."

"How often do you report in?"

Ire rose at the accusation edging his tone. "Every four hours. But I've only had time for one. Last night in the toilet."

"Christ." He tensed and turned partially away before facing her with a glare of anger. "There could've been a bloody shootout."

"Oh, come on. This isn't my fault. The phone is encrypted, and I doubt anyone way out here could intercept the satellite signal."

"Someone on the other end of your call could have reported your movements. The timing works."

"Wait a minute." She pointed at him, but stopped short of contact. "You think someone from *my* project office is responsible? That's crazy."

"Someone senior in this country ordered the CIO to snatch Mukono and Sipho. Whenever power and money are involved, so is betrayal."

"That logic seems pretty far-fetched." Although the reality left Joni questioning everyone involved in the operation.

"In this line of work, you have to consider all possibilities."

Her mind balked at accepting this one. "Lemmon gave me the damn phone."

"I take it you don't trust him?"

"Hell, no. He's lied to me about everything." Doubts rose almost as soon as she voiced the accusation. "Yet he secretively gave me your exact GPS coordinates when I flew in. If he was the traitor, he could have had Kagona strike when I landed. Does make me wonder if he didn't trust someone in the office. I thought his personal touch was being his usual clandestine self."

"He couldn't be positive I'd bring Sipho to the meet. Would the rest of the office know about your report?"

"Only the senior staff. But Lemmon answers the phone. It's possible he patches through to the control room. Still, the Americans and the South Africans have every reason to see that Sipho gets out alive."

"Greed knows no political boundaries."

"Or tribal ones." She shook her head, feeling hopeless. "This finger-pointing is getting us nowhere."

"It's laid out new obstacles we must consider going forward."

"Seems to me the biggest obstacle is what I haven't been told about this mission. No one goes to these lengths for bank accounts. And why didn't Sipho simply tell you where the information is hidden? Isn't it about time you level with me? You need an ally, Ian. You have to trust me."

"Ha. I can barely trust myself right now."

That's two of us, she wanted to say, but everything indicated this man cherished Sipho. If anyone had a champion, Ian was his. "Something else bothers me. Why would Brugman rush to snatch Sipho and risk pissing off the top enforcement guy in Zim? That would take balls."

"That's how Brugman operates. He's got his own power. His own men and network. Kagona uses it on occasion for his ends. That's what gives me hope we can get Sipho back. Brugman is going to make sure the handover happens in a secure place where he'll have control. He'll also want to negotiate price. For that, he'll want to gauge Kagona's desperation."

Joni leaned a shoulder against the wall and crossed her arms. "So you plan to waltz into a place where Brugman thinks he can hold off Kagona and ask for Sipho back? That sounds like suicide."

Ian gave her a wry smile. "Rather does, doesn't it. Got a better idea?"

"Give me a year and maybe I can come up with a plan. But you're right about one thing—it's obvious Sipho is worth more alive than dead to everyone. How much time do we have?"

"Enough. Everything isn't doom and gloom. I do have a plan. I'm not a novice at these types of things." Ian set a hand on the wall and hovered over Joni.

"I figured that out a while ago." She looked up at him and wished for her sister's height. Being short and slight seemed to be a drawback on this mission.

"What happened to Sipho is no one's fault. It is what it is." His demeanor had changed, almost relaxed, as though the thought of action gave him purpose. "All we can do is correct it. You can start by handing over the phone."

"You're kidding." She tried to straighten but his proximity prevented it.

"Not in the least."

His tone left no doubt he'd take it whether she agreed or not. "I'd like it back."

Without comment, he secured it in a pocket and stepped back. "We've a kid to rescue."

"I'm not a commando, no matter what Lemmon told you. I can't storm a fortress. You need to ask him for reinforcements."

"No time." He walked toward Sibongile's hut. "You said I need an ally, and you're right. You've intimate knowledge of the players in this game, and considering the size of our force, I could use help pulling off my plan."

"A minute ago you accused me of calling Kagona. Now I'm suddenly your ally?"

SANDY PARKS

He smirked. "Funny how life works."

"Look, I want Sipho back as much as you do. But I won't commit to anything without more details.

"Fair enough. We'll talk on the way to Bulawayo. Grab our stuff and let's go."

Sibongile sat outside her home with beadwork in progress spread on a cloth in front of her. They offered greetings but dodged any questions about Sipho.

Inside, Ian rolled up his mat and tightened up the fittings on his pack.

"You haven't said why we're going to Bulawayo," Joni whispered. "Is that where Brugman lives?"

Ian pressed a hushing finger to his lips. "No, he doesn't live there." He moved her out of the way. "We're going for something else." He reached down to a mat and with a quick flick tossed her the rolled skirt.

She caught it. "Bulawayo's a good-size city. I'll blend in better as a tourist." She laid the skirt on Sibongile's bed.

"Suit yourself." He scooped up his pack and slid it on a shoulder.

"Not so fast. What are we going for? Weapons?"

"Something that if I can't get, you can."

His implication sank in, and her insides riled. He was a chauvinistic SOB. No wonder he'd been so accommodating a minute ago. "Forget it. My womanly wiles won't get us anywhere."

To her surprise, he took her face in his large hands and pressed his mouth to her ear. "I may need you to steal a helo."

Shocked at his grand plan, she started to ask more questions, but he slanted his mouth across hers in a kiss. He released her as quickly as he'd entrapped her, leaving her distracted and trying to assess what had just happened.

Ian stepped outside and thanked Sibongile for the hospitality.

In a swirl of frustration, Joni scooped up Sipho's pack and

144

UNDER THE RADAR

scrambled after him. Once they'd rounded the hut, she fisted a hunk of his khaki shirtsleeve to slow him down.

"This whole situation stinks." When his brows raised, Joni rolled her eyes. "I mean with Sipho. What if the *sangoma* lied to the chief? What if Brugman didn't snatch him?"

"Got a better idea?"

She shook her head.

"Didn't think so."

With long strides, Ian started toward the chief's place, where he'd left his rifle. Ahead, the chief strolled to her vehicle, a worn Mercedes that had seen more than one generation of tribal leaders. They headed toward her.

"What about your rifle?" Joni asked as they passed the chief's house.

"I can't take it into the city. We'll stand out enough as it is."

"Won't we need it to go after Brugman?"

Ian looked at her, surely seeing the terror building at the idea of jumping into a pit with a bunch of gun-toting bad guys. Yeah, she was scared and willing to admit it.

He slid an arm around her shoulder and squeezed. "We're not going in unarmed, but it hardly matters. The guy has an indecent amount of money at his disposal. We'll need brains, not weapons, to outfox him. And that, if you're willing to go along, is where we have the biggest advantage and opportunity to succeed."

145

CHAPTER FOURTEEN

Ian opened a rear door of the battered vehicle. They had a logistics problem.

The chief grimaced. "I have load for Bulawayo market." Baskets overflowing with beaded dolls and jewelry, carved elephants, and sculpted stone figurines covered the seats front and back. "Fuel is hard to find. Each trip I must take much. Sit on floor. At roadblock, cover up with blanket and baskets."

Chief Zwide took a grave risk hauling them to Bulawayo. If discovered, she could end up in jail, an unpleasant place where many did not survive.

"How far to town?" Joni asked her.

"With one roadblock, maybe one hour. Longer if more."

Ian moved baskets and shoved the front passenger seat full forward. He squeezed onto the rear floor with his legs unavoidably sticking across to the opposite door.

Joni shut the door behind him. From the other side, she made several attempts to sit facing him on the back floor. Her feet became dangerous weapons with no place to go.

"Quit the gyrations." Ian caught her foot. "Turn around before I get booted and sing like a bird." He motioned her to sit on his lap and lean back.

She rested her butt atop the center hump running down the floorboard and lay awkwardly back against Ian. His pack went under her feet and she clutched Sipho's tightly between her knees.

The chief piled baskets back in and shut the door. She drove down the dirt road with windows open and dust swirling. The jostles and bumps tossed Joni about until Ian dragged her farther onto his lap and against his chest.

"You know, Taljaard, when it comes to women and transportation, you have a real problem."

"That so? Doesn't seem so bad to me. I've a doubting woman in my grasp, wood carvings jabbing me in the side, and the smell of exhaust when I bury my face in your hair." He nuzzled close to her ear and whispered, "Can't think of anything else more likely to get a rise out of a man."

A hard jab with her elbow arrested any possibility of that occurring. For the next fifteen minutes the dirt road and its accompanying jarring caused the Mercedes as well as his body to creak and moan. The need for oxygen among the exhaust and dust made talking near impossible. The chief poked the radio on and music added to the din.

A turn onto asphalt brought stability, fresh air, and cool relief. Sweat stuck the back of Joni's shirt to Ian's clothes and chest. She leaned up to cool off. Her weight shifted.

He groaned and pulled her back, resting her head against his chest and shoulder. "You're going to have to stay still or I might not be in shape to rescue Sipho."

Joni fingered the strap of the school pack. "Since I have your undivided attention, tell me about Sipho." She positioned her mouth near his ear and kept her voice soft so it wouldn't carry to the chief over the music. "What does he know about the information?"

"You saw what he did with the chessboard. He remembers details. Like shapes and patterns, names and numbers."

She stiffened as the truth sank in. "Oh, my God. He really is the information. Are you sure he still remembers it?"

"At times he spouts off names, numbers, and banking institutions. I remember what I can and pass those along. They've checked out. I've tried to get Sipho to recite the content of the papers he saw on his father's desk, but he tenses up. The details come out when he's relaxed or whenever he chooses."

"Lemmon probably isn't happy about that. There's no guarantee the Americans can get the information out of Sipho, either."

"Lemmon claims he has a psychologist standing by."

"He's a bottom-line kind of guy. I wouldn't trust him to do what's right."

"But I trust you will. And in the long run, it isn't all about what Sipho knows. It's keeping the knowledge of what Mukono might have told us out of Kagona's hands." To emphasize his personal faith in her abilities, he brushed a reassuring hand along her arm. At a time when things in his life were falling apart, he found holding another human comforting. He not only wanted but needed to believe she had no part in Sipho's kidnapping.

"Lemmon and I don't exactly have a good working relationship." She shifted uncomfortably and he suspected her precarious position wasn't the reason. "To put it mildly, we have issues."

"Is that why he didn't brief you on Sipho?"

"To be honest, if I had known about Sipho, I wouldn't have come. Lemmon knew that."

"What do you have against kids?"

She bit her lip, likely debating how much to share. "Last time I did something for Lemmon, a kid got killed. I was flying for the military at the time. Lost two other people on that mission, but the kid sticks with me. It's a hard thing to forget."

That's one more reason why he had to save Sipho. "Was the mission in Colombia?"

148

Joni gave a little snort. "That obvious, huh? The details won't help this situation any, if that's what you're thinking."

"I'm thinking I want to know what makes my team member tick." Not sure why, he wound a single tress of her hair around a finger. "If you're going to melt down at something, I'd like to be forewarned."

She tugged the strand from his hand. "Meltdown is not a problem. Tell me more about the Mukono family."

Her reaction triggered a light chuckle from him. "Give a little to get a little. They trained you well. I shouldn't have expected less. Only it appears I'm doing most of the telling." He shifted, hoping to restore blood flow to his lower body. "What else do you want to know?"

"How did you end up with Sipho?"

Ian rested his cheek against her head. "The day of the accident, I'd come up to Harare. Sipho and I spent the day watching the zebras and giraffes at the Mukuvisi Centre. That night, the whole family was invited to a dinner. Sipho had a meltdown, so they called me to discreetly come get him. We went out to watch stars until he settled down and came out of his shell. I figured it was time to get him home. His father and I still had business to discuss that night. When I turned onto the road to his house, something seemed off. A government vehicle parked down the street as a lookout, and a strange car sat in front of the house. No one recognized my truck, so I kept on driving."

"Smart move." She wiped sweat off her brow. "I'd probably have gone inside to see if something had happened."

"Nomathemba said she'd call when they got home if we weren't there. I hadn't heard from her, but cell phone service here can be erratic at times. Still, we all knew the risk involved in investigating anyone in the reigning political party, in particular those wanting to take or hold onto the president's job. My suspicions were raised. Reports came out the next morning, claiming a drunken Mukono had killed his wife and split" town with millions. Sipho was the only end they hadn't tied up."

A basket vibrated off the seat onto Joni's lap. She shoved it back up. "Mukono's not the first person a government has thrown in prison with a trumped-up cover story."

Ian nodded. "Later that night after I'd secured Sipho somewhere safe, I went back to see what Kagona had done in the house."

"I take it the CIO didn't find the list."

"That's why Sipho's father is still alive. He had an incendiary device inside the safe in which he kept discs and documents. When Kagona's people tried to crack it, the information disintegrated. They've no idea how deep he dug into their finances or which accounts and assets might have been compromised."

"What about his computer?"

"He intentionally kept the list on paper and used various computers for research. That's one of the reasons he required help in tracing down the accounts. By Kagona's actions, I assume he hasn't been able to compile a decent track of the minister's research history online."

"A son would make a good tool for getting his father to talk. How did they discover Sipho had the list in his head?"

"My guess, from the security camera in Mukono's home study. The machine was gone when I got there. Kagona likely saw Sipho looking at papers on his father's desk."

"Yeah, but I could look at papers, too, and remember little."

"The camera records for about four days before the files are destroyed. Three days before I arrived, Sipho had seen a list of names and addresses on his father's desk and challenged his father to a game. Whoever could recite the most names and numbers won. Mukono told me about it the morning I arrived. Then he was quickly attempting to delete a newer entry on the tapes the morning I arrived before he had to leave for work."

"What was he trying to erase?"

"The night before, Mukono was studying the data he'd been collecting, looking for connections. He paused for a toilet break but didn't lock up his papers. Nomathemba distracted him briefly to talk and ask if he wanted more tea. He returned to find Sipho sitting in his chair, staring at the papers. He'd had a good ten minutes to look at them. This time Mukono was more concerned than amused. He asked Sipho to recite a few of the accounts, and he repeated them exactly. Mukono's hands shook when he told me about it."

"Did Mukono erase the first incident, too?"

"He didn't say, and I doubt he had time. He may have figured it would be copied over the next day and not bothered. But he did tell Sipho the game had to be a secret between them. That's part of the reason he is reluctant to share what's in his head."

"Poor kid," Joni said. "If Kagona saw that part, it would expose his capabilities."

"*Yebo.* And that scares the *scata* out of me."

Slowly, as though letting this news digest, she rubbed her palm across the top of his hand on her abdomen. The sensation added another layer of complexity to his screwed-up insides. "Mukono and Sipho are lucky to have you as a friend. I understand your fear. If either Sipho or his father gives Kagona the information, he'll have what he needs for damage control."

"And he'll kill them both."

She squeezed his hand. "We'll get to Sipho first."

He relaxed his grip, having unintentionally clenched her tight while thinking of the mess. "Care to tell me more about Colombia?"

"I'm going to prop myself up for a moment and give you a chance to breathe." She hooked one hand on the front seat and the other on the back and lifted slightly.

Ian assumed she had a good deal to tell, but in a business where people rarely revealed all the truth, he doubted she would divulge much more. When she trusted him enough to tell all, perhaps he'd consider revealing the more threatening reason for this mission.

Seconds later the chief tapped Joni's hand. The radio went silent.

"Roadblock ahead."

Joni dropped back on Ian's lap and shifted a woven blanket and several stacked baskets over them.

"How many vehicles in front of us?" he asked the chief.

"Five, and four men checking. They are taking one truck off to side. Two men now are looking at cars."

He heard commanding voices and engines starting, stopping, and idling. They rolled closer to the checkpoint. The air grew stifling as they huddled under the baskets with no incoming breeze and no air-conditioning.

"One more," the chief reported. She caught her breath. "It's Jabu."

"The brigade kid from the kraal?" Joni asked. "Will he stop you?"

"When I teach, he let me go. Today, I don't know."

"What's left in Mrs. Lubbe's snack bag?" Joni whispered to Ian.

"A rice cake and one cookie."

"Lift up a bit on the baskets." She clawed at Sipho's pack, which had slipped from her lap. Her fingers snagged the strap. She dragged it closer and slipped her hand inside.

"This is no good." The chief had worry in her voice. "Three cars arriving fast."

Ian pushed Joni aside and leaned up. He peered out from among the crafts. "An envoy." He sank back to the floor, pulling her with him. "It's the CIO."

CHAPTER FIFTEEN

The car pulled forward as Joni slipped the cookie through to the front seat. The chief snatched it and greeted Jabu. They spoke for a moment.

Joni held her breath, listening to her heart beat loudly enough to play to an audience. As the car began to move, she relaxed and sucked in new air. Somebody yelled in the background. She didn't have to speak an African language to recognize "stop."

The chief continued to accelerate slowly as though she had nothing to hide. No shots rang out or vehicle wheels spun in hot pursuit. As voices grew more distant, Joni asked what happened.

"I said to Jabu too bad he missed singing at *pungwe*. I gave him cookie. He waved me through as men jumped out with guns and yelled to stop cars. Since I already was going, Jabu stopped car behind me."

Ian helped Joni shove the baskets up and back onto the seat. "What did the men look like?"

"If I looked too much, they might try to stop me. I only saw blue uniforms and big guns."

Ian sat up taller and settled Joni back on the central hump. "You did the right thing. Any more roadblocks up ahead?"

"No more. I will drop you in the city. Tell me where."

"Drop me at Josiah and Eighth Avenue, and Joni at Centenary Park."

Joni shifted around to look at Ian. He stared back, his face betraying nothing.

"Chief Zwide," Ian said, "can you please find something on the radio again?"

The chief seemed to understand their need and hiked the volume.

"Why are we splitting up?" Joni whispered, suspicious he might have plans that didn't include her.

"Bulawayo is my stomping ground. A number of my friends and associates live there. Kagona is aware of that fact. My contact would not appreciate me exposing him to our troubles unnecessarily. I can only hope he will choose to help."

"Is he the guy with a helo?"

"That's him. I'll make myself visible where he hangs out and then we'll meet in a safe place. If he saw you with me, he wouldn't make contact."

"So why send me to the park?"

"At the moment, out in the open among the tourists is probably as safe a place as any. I shouldn't be long and will come collect you as soon as possible. If I think the park isn't safe for a meet, I'll send someone else."

Another fifteen minutes and a smattering of additional details later, the chief turned down the wide tree-lined streets of old colonial Bulawayo. She dropped Ian off first, and he headed toward a hotel in town where he intimated his contact hung out. He didn't have time to elaborate and Joni didn't ask.

Her mission was merely to wander around in a nearby park like any tourist and waste time. If Ian failed to appear within thirty minutes, she was to catch a ride on something called the Lady Morag at King Arthur's Cross station.

The chief tossed Joni a light woven scarf to throw around her neck to disguise the blemishes marring her shirt. After bidding the chief farewell a few blocks later, Joni got out at Centenary Park. Unlike parks with extensive green lawns where buildings peeked through the shade trees, this one had rather a mixture of fountains, lawns, shrubs, trees, and wild bush areas. The years had taken their toll, and some parts were marked for restoration.

Chief Zwide's scarf covered the taped tear and bloodstains on her shirt. She tucked her red hair up under Ian's Aussie-style hat on the chance her description might be making the rounds. In all likelihood, Kagona had posted alerts in Bulawayo after getting the details from his helicopter crew.

Doing her best to look interested in the surroundings, she strolled into the park. At any other time, she might have enjoyed the low-key, relaxed atmosphere of Bulawayo, but not with a life at stake, and an innocent one at that.

She had never been particularly good at waiting. Unable to control the urge to check the time every minute, she stuffed her watch hand into a pocket. Around her people milled and chatted, some heading to the park for an early lunch, others leaving after a late break. No one seemed inordinately curious about her presence.

City office workers, black and white and a variety of nationalities from Asia and the Middle East, strolled the grounds. Their dresses or business-casual wear was in sharp contrast to the beat-the-heat shorts the few tourists wore. She aimed for the tourists.

Not knowing exactly what Ian meant by riding Lady Morag, she scouted out the park, wondering if perhaps a stable lay hidden in its interior. A sign to King Arthur's Cross soon caught her eye. Children's voices full of shouts and giggles led her to a playground. Near it, a half-scale watering tank had a placard with "King Arthur's Cross" on it.

Evidently the park had an amusement train ride. Signs

bragged it had recently been refurbished. She walked along a decorative fence to a train turntable and engine shed. Next to them sat a sky-blue miniature train on a narrow track. It pulled an open carriage wide enough to hold one adult or two small kids on each of a dozen benches. The name on the engine read "Lady Morag." She smiled.

The tracks wove around and through a putt-putt golf course and disappeared into an area of wild bush and grass. Not a fast escape route if things went to hell, and the chances were high they would, but she'd make do.

An itch niggled at her neck as though someone were watching. She shaded her eyes with her hand and pretended to follow the call of a bird. She scanned the park. If someone had an interest in her, they blended in well.

Three girls lined up for a ride with their mother, while two boys talked with a balding white guy about her dad's age. The man strolled with a hitch in his step past her and up to the train with the boys traipsing close behind. She checked her watch. Thirty-two minutes had passed. She had waited long enough. She dug out a crisp US dollar bill Ian had given her, the *new* currency since Zimbabwe's had failed with superinflation. Next time she planned to fly with emergency currency tucked in a pocket.

The engineer took her money and told her to get in line. When he signaled all aboard, the girls scrambled to the front. The older guy worked his way to the last bench, so she picked one in the middle. The young boys took seats behind her.

The little steam engine hissed and puffed as it got under way, tugging the carriage car at a mind-numbingly slow pace. Kids chatted and laughed. The boy behind her bumped against the seat and placed a tiny walkie-talkie next to her. She casually scooped it up and set her elbow on the seat back. With the walkie-talkie palmed, she placed it against her ear.

A voice said, "Don't turn around, Miss Bell."

He knew her name. Likely the old guy in the back.

"You should have put Ian's hat in the pack," the voice continued. "It's a dead giveaway. Now, when I tell you, exit the train. Step over a little fence and head through the park. Go west along Leopold. You'll see a hostel called Irvine Place. Go in and then out a rear door. Hop into a white SUV and keep your head down."

"And why should I trust you?" She rubbed a finger across her upper lip while speaking.

"Because there is a man, with a maroon shirt, waiting for you at the station. He's not a friend of you or Taz."

The mention of Taz infused adrenaline into her system. Still, Kagona could know about Taz. "More."

"Christ, missy. Ian said the boy sang better than you."

"Good enough." She shifted her feet farther apart, ready for a quick departure, scanned the area ahead, tightened her grip on the pack, and, shit...noticed her bootlace untied. Like a madwoman, she leaned forward and grappled with the shoestrings.

Ahead the tall chain-link fence holding back tall grasses and greenery wanting to encroach on the track disappeared, and a short green metal fence appeared.

"I think you can manage this. Off with you, girl."

Her fingers fumbled with the bow.

"Now, girl, now."

She yanked the loops snug, scooped the walkie-talkie into her shorts pocket, and hopped off the carriage, much to the chagrin of the engineer, who kept track of his passengers with a rearview mirror. In two steps, she hurdled the fence and accompanying bush and briskly walked across the park, giving the station area a wide berth.

The temptation to look back at the miniature railroad and see what happened to the man on the train or the one waiting nearly overwhelmed her. The walkie-talkie remained silent. She kept to the grass and park trails among the trees away from the roads until the park ended at the intersection of Samuel Parirenyatwa and Leopold Takawira. A guess said this street was the Leopold the old guy meant.

She stopped and scanned traffic on both roads, using the opportunity to look back at the park. Walking on a path back into a wooded area was a man in maroon. He hadn't walked past her, so he must have been following and turned around once she stopped. He must have picked her up after her disappearance from the train.

As a cold chill took hold, so did the urge for action. After a sedan whizzed past, she hustled across the street and up the block. The Irvine Place hostel sat back from the street. She cut toward it, casting another look. No trace of the man. She entered an inside corridor with chairs outside room doors. After a quick stroll past pink colonial columns matching the town's theme, she exited through a back door.

Outside a battered SUV sat backward in a spot ready for a speedy getaway. She opened a back door on the passenger side and stretched her body across the rear seats.

Ian sat up in the driver's seat. "Took your bloody time." He started up the vehicle and then pulled out onto the street.

"It's a big park." She pulled off his hat and shook out her hair.

"Stay down. Your red hair sticks out like a beacon. You ever thought of making it a more common color?"

"I didn't plan on skulking about Zimbabwe." A wheel hit a pothole, but he didn't slow. After two turns she asked, "Where are we going?"

"Around the block for the moment." He had barely finished talking when he shoved on the brakes, rolling her off the backseat and slamming her against the front ones.

"How about warning me next time?" Before she could pick herself up, the passenger door opened and the man from the train jumped in. Ian hit the accelerator, tossing her back on the seat.

She propped her head up with her hand as she lounged on her side. "Who taught you to drive?"

Ian grinned over at their visitor, a rather uniquely dressed individual with his khaki pants tucked into high-topped boots and a short-sleeved black T-shirt covered by a sleeveless leather vest. A portion of a tattoo stood out below the short sleeve. Motorcycles and Harleys jumped to her mind.

"Joni, meet Mad Mike," Ian said. "Mike, Joni."

"We met," she mumbled, wondering how many more people would get involved in this *simple* mission before the day was over. The project office had expected her back in South Africa eighteen hours ago with the fact Taz had ever landed in Zimbabwe a secret.

"And a pleasure it was." He extended a hand with oil-stained nails back over the seat. Roughened skin slid across hers as he firmly clasped her hand. For an old guy with a less than stable gait, he seemed pretty self-assured. Frequent sun exposure of his neck and face had thickened his skin. The rich tan carved years off his age, but sun damage added wrinkles. She guessed the newcomer well into his sixties.

She sure hoped Ian hadn't rounded up some nutcase believing they were out of options. The name Mad Mike had an ominous ring. That, combined with his ponytail gathered from what little hair he had left, and a tiny diamond stud in his left ear, labeled him as a free spirit. Seemed like a great combination for disaster. She handed him back the walkie-talkie.

Ian cast Mike some sort of unspoken question that smacked of tight familiarity.

Mad Mike nodded. "He showed up moments ahead of me. A local, but CIO. Probably tipped off by the police."

"I saw the man, too." Joni figured Ian would want to know she'd been followed. Of course, he could have asked her directly. "He was dressed in maroon and pretending to walk back into the park right before I crossed the street. He hadn't reappeared when I turned for the hostel."

Mad Mike smiled with a pleased gleam in his eye. "That's because he's snoozing like a big cat under a tree."

Ian frowned and concentrated on something outside. "How much time do you think we have before they connect us up?"

"Thirty minutes tops," Mike said. "There're only so many ways to get out of town. Kagona will figure you'll go for the fastest."

"Did the maroon guy see you before you popped him?"

"Never knew what hit him. If the CIO show up later, I'll say you hired me for a trip to Vic Falls."

"And if they don't believe you?"

"Hey, there's always Zambia."

"Whoa." Joni put a halt to their running conversation. "Why do you think Brugman and Sipho are in Victoria Falls?"

Ian shrugged. "He has a big lodge there. It's my best guess as to where he's taken Sipho."

"That's a long way to go. Why not hold him locally?"

"If he's planning on selling Sipho to Kagona, he'll want to be in secure territory so Kagona can't easily retrieve him. His lodge is the perfect place." Ian looked back at Mad Mike. "We can steal your helo if taking us puts you in a tight spot. The little lady back there's a pilot."

Mad Mike studied her with more scrutiny than before. "Really? I'll be damned."

Ian snickered. He turned off a major street to a narrow one, taking them away from the city center. "So, you game for us to take the helo?"

"And leave me behind? No one would buy the story. Blithering idiots have tried to steal her before. They won't be back. Kagona knows you don't fly, and no way would he suspect this pretty young thing could get it off the ground."

Personally, this pretty young thing didn't care to be in the same chopper with Mad Mike. The guy lived with his rotors an inch from the hillside. She preferred not to be around when he crashed and burned. "What did you do to the guy in the park?"

"You don't want to know," Ian and Mad Mike said in unison.

Mike socked Ian in the arm and shook his head as though clearing forty years of cobwebs. "Sure miss the good old days, kid. Your dad was the best."

Best at what? No mention of Ian's father—other than his owning a ranch—had been in his dossier. Was he some nutcase adventurer flying around with Mike? Joni shook her own head. Hope of rescuing Sipho deteriorated by the minute.

Ian's grip on the steering wheel tightened at the mention of his father. "You aren't taking on his kinda fun on the side, are you?"

"Here and there. You never get it out of your blood. Been slow working around here since he left. Not as many animals to see from the air, so fewer tourists who want to fly. Lately my biggest excitement is soaking passengers in the mist on occasional flights over to Vic Falls. Hell, taking a flight from Bulawayo to there is damned expensive. The only positive is my helo is in better shape than the train. You know Zim's on all the tourist warning lists."

Ian glanced back over his shoulder at her uncomfortable position. "Only a few more blocks before you can get up."

She forced herself not to peek up over the seat. "Do you guys play this game often?"

Ian cast Mike a look, which said to quash any stories he might be dying to tell. "Occasions have arisen."

In a disintegrating country like Zimbabwe, she could only imagine the occasions. "How far out of town is your chopper?"

"I park her close in, 'cause I can't get petrol to drive

customers out to the airport. Jet fuel is particularly hard to come by but obtainable as long as I bring in tourist dollars. Had to get a damn party card for any type of fuel I can secure."

"I take it you're talking political, not booze and fun?"

"*Yebo.* No card—no fuel of any kind. Damned way to make a living."

"You could leave," Ian said halfheartedly as though knowing the answer.

"Don't see you packing up shop." Mad Mike pointed Ian down another road. "We need to stock up."

No surprise there. Joni would bet he didn't leave a six-pack far behind. "I don't have much of an appetite," she commented.

"Ha." Mad Mike slapped his knee and grinned over at Ian. "I didn't think Americans came with a sense of humor. Who sent this one?"

If there was one thing Joni detested, it was being spoken about, not at. She sat up on the backseat. "Perhaps contacting Mike was a lousy idea." She tapped Ian on the shoulder. "Dump him out and we'll find his chopper on our own."

"Aw, now you've gone and gotten all American on me." Mad Mike frowned and gave a weak attempt at looking hurt.

"Naw, she's a trouper." Ian tried to smooth feathers. "Just been screwed over by management. Seems they didn't give her all the details."

"What does she know?"

"She's figured out the information about our fearless leaders' assets is in Sipho's head."

"Oh," Mike said, as though she was about to be screwed over one more time.

Had Ian spilled his guts to Mike about further details, of which she was still in the dark? "I'm willing to expand my knowledge base any time. Feel free to fill me in."

"When the time is right." Ian didn't sound at all sincere.

She crossed her arms, irritated but also curious as to what

else was going on with Sipho and why she wasn't on the need-to-know list.

Ian pulled up to a quaint little house with dirt for grass in a ten-by-ten front yard. Mike's place wasn't all bad, though. Two huge clay pots held purple and red flowers. A metal roof sheltered a whitewashed wood frame. Not exactly the Harley garage she'd expected of Mad Mike's persona.

"Wait here," Ian said as he reached back and grabbed his pack. He hopped out to join Mike. Instead of entering the house, they cut around back out of sight. There was something funny in Mike's stride, like she'd seen in the park. She waited with folded arms, curious to see what they retrieved.

Eventually Ian appeared alone and dumped two heavy packs in the rear. He took the passenger's seat this time, leaving the driver's seat open for Mike.

"He's crazy, you know." She felt rather stupid pointing out the obvious. Of course, Ian might be so caught up in the old boy reunion he didn't notice.

"Yep. Known that for years."

"You made contact at a hotel. Was it in the bar?"

"That's where Mike drums up business."

"And how many beers did you see on his table?"

Ian frowned, but humored her with an answer. "I didn't go into the bar, and it doesn't matter. He could fly a helo dead drunk on a stormy day. He's good."

"Then you're crazy to fly with him."

"Don't get me wrong. Mike's no fool. He likes his Zambezi lagers, but he isn't going to jeopardize business by drinking. He sits in the bar because it's in view of the lobby and prospective clients. With Sipho's life at stake, I wouldn't let him fly drunk."

"I take it you've noticed his walk?"

"It's a bad hip."

"And he's still able to fly?"

"You bet. He's the best around."

Joni cringed at the male overconfidence he spewed. "From what I've seen of Zimbabwean pilots so far, that's not saying much."

"You'd be surprised."

"It's not good to be surprised in a helicopter. People die that way."

"You always a pessimist?"

"I call 'em how I see 'em. You always an optimist?"

Ian shrugged. "What else can you be in Zimbabwe and still want to stay?"

"Look. Don't risk our lives or the chance to save Sipho. Let me fly the damn thing."

"How much time do you have in a Jet Ranger?"

"My career is based on flying and testing different helicopters. I've twelve different ones under my belt."

Ian raised doubtful brows.

"Part of my job is to quickly get comfortable with any make and model. I can do it."

"Time in the Ranger?" he asked again.

She sighed and glanced up at billowing clouds in the sky. "Four hours."

"I think I'll stick with Mad Mike."

"Damn it, Taljaard. I've several *thousand* hours in choppers. This is my job."

"Mike has seven thousand, and he's saved a helluva lot of people in his time. I've seen what he can do."

As though on cue, Mad Mike came out of the house with a pack on one shoulder and a bag in hand. He tossed them in and slid behind the wheel. "She's parked right around the corner. We should be in the air in minutes."

Minutes left little time for a preflight inspection and fuel check. Not only whether they had any fuel, but also whether deadly moisture had condensed in the tanks. She wanted to make sure grime hadn't infiltrated any of the systems and that the maintenance was current. No reason to risk their lives only to have some half-baked pilot kill them all with negligence.

"What model do you fly?" she asked, still trying to determine whether he normally acted so brash or had downed a few too many Zambezi lagers.

"She's the sexiest 206 Bell Ranger in Zim. Good handling characteristics and visibility. The tourists love it when I buzz white-water rafters on the Zambezi."

Impressive. The more he talked, the more determined she became to avoid putting one toe into his chopper unless she was at the controls.

The road dwindled to a rough one lane.

"There she is." Mad Mike slowed the vehicle and pointed proudly off to the right.

Joni stared, agape. It wasn't a question of who would do the flying. That thing would never get off the ground, especially not in the minutes Mike had bragged. Rusted red best described the paint scheme of his pride and joy. A piece of engine cowling leaned up against the dented pilot's door, and debris hung from the rotor. No self-respecting tourist would risk their life in that machine. Nor any pilot she knew. They'd need to find some other way to save Sipho.

CHAPTER SIXTEEN

Mad Mike burst into laughter.

"Did you see her face?" He punched Ian in the arm. "Works every time."

"It's not nice to piss off a woman, Mike. Particularly one who has the smarts to get even."

Joni examined the chopper a little closer, as they approached. Mike parked near some trees and brush, which they used to hide the SUV. Damn, he'd done a nice job of camouflage.

Not waiting for the men, she walked over to the helo. The sign on the door read "Adventure Tours." No kidding.

Mike rambled up. "I use the makeup to keep the plunderers away. Can't watch over her twenty-four-seven. Works pretty damn well." He slid a sock-like cover off the nose and front windows, then peeled off plastic sheets of rust, dents, scrapes, and dirt that clung by static electricity to the fuselage.

"No bullet holes?" she chided.

"Nope. Afraid it might encourage someone to add to them." Once he pulled off the last piece of fake debris, he stepped back. "I always say a woman looks good in red."

Polished and with little sign of wear and tear, the red-and-white Ranger did gleam. Mike might be mad, but the attention to his equipment earned him a few points of her respect, especially living under tough circumstances.

Ian loaded the packs into the Ranger's compartment while she and Mike did a walk-around inspection and removed any remaining pieces of fake debris. Mad Mike took one side and she took the other. She checked the condition of the main rotor blade, landing skid, and a half dozen other helo parts before meeting Mike at the tail boom.

When they met halfway, he accepted without question she had satisfactorily completed her half. The fact she was a woman didn't appear to produce any doubts about her ability. Perhaps he did have some redeeming qualities after all, or perhaps he didn't give a damn about the walk-around and would have skipped it completely if she hadn't been present.

Before they boarded, Mike and Ian loaded jerricans of fuel they had hauled from a multiply padlocked shed. They centered the heavy, awkward cans in the back seating area and fastened them down. At least Mad Mike had some sense of the delicate nature of center of gravity.

Against her better judgment and ready to take over control at the first sign of trouble, she piled onboard. A not-so-refreshing scent of fuel filled the air. Mad Mike took the right command seat and she took the left. Both seats had full controls—a fact she kept close in mind as she fastened her harness.

Ian tightened the lids on the fuel cans and took the back. He slipped on a headset with ear covers for sound suppression and adjusted the mike near his mouth.

She fastened her harness before donning her headset. "How far to Victoria Falls?"

"About four hundred kilometers." Mike glanced away from a slim checklist hooked by his leg. He flipped on an overhead battery switch, adding a beeping from the caution panel to the confined space. "That's two hours in this gal."

By habit, she checked the full travel of the cyclic stick between her knees. Mike set switches overhead, checked instruments, and started up the engine. The whine grew louder and louder. Sunlight and shadow flashed when the main rotor began to turn. She scanned the temperature and pressure gauges.

He gave it a one minute run-up, checked hydraulics, and then shot her a sideways grin. They lifted off. The chopper rotated right then left. Mike overfinessed the anti-torque pedals. The average tourists would simply assume the spin part of the adventure. A sixth sense hinted he had purposely done that to test her reaction.

"It's my bad hip." Mike grinned rather gleefully at the tense face she must have presented. "Just wait till you see what else I can do."

"Save the theatrics for the tourists. Show me you can fly this thing."

"No problem, missy. Just sit back and enjoy the ride."

He aligned the nose into the wind, adjusted the cyclic pitch control, and pulled up on the collective. At an altitude clear of surrounding obstacles, he added power and transitioned to flight, all with smooth, well-practiced precision. Bad hip? Right. He worked the anti-torque pedals like a pro.

Ian handed over a hunk of cheese from the rear seat in a clear attempt to distract her. "Lunch?"

Even though her stomach had long ceased its feeding growls and moved to harvesting what little fat she had, the sight set her mouth salivating. She twisted to reach the cheese. Ian held onto the piece, forcing a brief tug-of-war, before he let go. A single dimple deepened in apparent amusement.

"Do I have to wrestle for bread, too?" she asked.

He waved a piece and handed it over. "You need to relax, distract your mind, or you'll be exhausted before we get there."

"That's hard to do when you claim Sipho is in the hands of a man you rate next to the devil. Not to mention you suspect he'll be sold to a man not above torturing him to get information."

The tiny muscles along Ian's jaw perceptively tightened. "Thinking the worst going into an operation will only lead to failure. Getting Sipho back will not be easy, but Mike and I have done this type of mission before, and I suspect we'll do it again."

"Then tell me about your game plan. You do have one?"

"Your faith in me is staggering."

She loosened the harness and turned in the seat to better see into the back. "I could say the same of you."

"Fair enough." He handed a chunk of cheese up to Mad Mike before scrounging in one of the packs and extracting a small plastic marker board.

After a minute of sketching, he passed up a drawing. Joni held it where Ian could reach from the backseat and point out features. "We'll take a closer look once we land, but this will give you something to think about. Brugman lives in a safari-type lodge set back from the Zambezi River above the falls."

"And what a spread," Mike interrupted. "A ruddy resort all done up like an African village. Geometric designs. Thatched roofs. Stucco walls made up like mud. Nature park on one side and the Zambezi River on the other. Only a few special visitors carting large amounts of cash can partake."

Ian pointed back to the marker board. "Don't be fooled by the looks. Brugman has money and the hottest techno toys. His men are well armed and equipped with night vision. He'll have perimeter alarms and cameras, and his property is inaccessible except by river or air. There is a dirt road in the national park, but we can't get a vehicle in close enough because it is heavily monitored due to poaching, and it has one entrance and exit. To discourage any encroachment on his private resort, Brugman leaves traps for unwanted visitors around his property. He also monitors the river shore within thirty meters of his dock. If he has any guests, like Kagona, they'll most likely arrive via helo and land in a clearing north of the main lodge."

Joni stared dumbfounded at his crude map. "And you plan to take on this complex alone? I assume you have a Special Forces squad and machine guns hidden in those small packs."

Ian cocked an innocent brow.

"None? Figures. How about grenades?"

He shook his head.

"Night-vision goggles, since it'll be dark?"

"I helped there," Mad Mike responded. "They were on my equipment list."

"How about communications gear so I can hear the sounds of you being drawn and quartered?"

Ian added a smile, and Mad Mike chuckled. "She reads you pretty well, kid."

"I have what we need," Ian said. "We'll reach the villa by boat, putting in downstream. You can get a boat, Mike?"

"Transportation's my game."

"Good. Once on land, we'll split up and I'll head to the lodge."

"And you plan to walk right in and ask to see Sipho?"

"No need. I know where he's held."

"Are you psychic, or have you been there before?"

Mike wrenched his neck around to grin at Ian. "Gonna tell her about that one?"

"Nope." He slid a hand to the back of her seat and pulled himself closer as though to emphasize his control of the situation. "If it makes you feel any better, both Mike and I are intimate with the place."

"Well, I'm not."

"That's why you're staying with Mike. He'll be making sure they can't follow us out by boat. I'd leave you with the Jet Ranger, but in case Kagona flies in on a helo, Mike might need some help to take care of the boat and the helo."

"At last, part of the plan that makes sense. I can evade a few armed guards at a con man's villa. I've done it for weeks before in a jungle."

"She's right about that, kid. With my *hip* and all, I'll need the help."

"Don't feed me the crap about a bad hip." Joni had tired of their guy games. "I've known enough chopper types who came back from war with injuries. A good number returned to flying. Is it your foot? Leg?"

"Foot."

"Gone?"

"You don't mince words, do you, missy. Lost it back in the Rhodesian conflict."

"It seems Joni reads you well, too, Mike," Ian said.

"So why the hip story?" she asked.

Mike shrugged. "Habit. Tourists are edgy enough about flying without knowing their pilot's missing a bloody foot."

"I can believe that. But I'll be the one to knock Kagona's chopper out of commission if it's there. You two have made Sipho my responsibility, and it's time I accept it."

"A bit touchy, I'd say." Mad Mike looked over his shoulder at Ian and then at her with some misplaced sympathy. "Sounds like you trusted someone once to do a job and they let you down."

"I've no one to blame but myself."

"Heard that before." Mike grinned. "Did this job have anything to do with that jungle trek you mentioned a minute ago?"

"You boys have your secrets, I'll have mine."

"Not when I'm running the show," Ian cut in, dead serious.

And that's exactly what this adventure promised to produce, a one-act play destined to see the curtain drop quickly. "How did wildlife ranching and chasing poachers make you an expert in combat tactics?"

"Humph." Mike cast a curious look at her. "The CIA must

have missed a few facts in your briefing, missy. I can see why you two are having communication problems."

She locked her focus on Ian. "Exactly what have I missed?"

"You tell me more about why they chose you for the Colombia mission, and I'll tell you about Mike and my father. Then perhaps you'll understand why I can pull this off."

"I know better. You first."

"Ha, she's got you figured."

"Quit being a smart-ass, Mike, and fly." Ian sat back and set a hand on each knee, staring at his fingers as they drummed his leg. "This maniac flying the helo and my father trained and operated as Selous Scouts."

He paused for her reaction. If he expected shock and awe, she gave him none. Everyone in Africa had heard of the elite commandos famous in Rhodesia before independence renamed it Zimbabwe in 1980. In the last thirty-five years, the men had turned into mercenaries for hire or faded away. Now most had retired. Just because the father had learned every survival trick on the continent and could track and destroy an enemy didn't mean his son had the same capabilities.

"My father is a retired surgeon, but I wouldn't trust myself with a scalpel."

"I get your point. But he probably never spent his every moment with you in training. Mine did, quite to the detriment of our relationship. As soon as I returned from university, he disappeared to where he claimed they needed him. I ran the ranch and helped develop the conservancy until Mike came knocking one day. Said he needed help to get my father's arse out of trouble."

"Ah, 'twas the start of a sweet relationship." Mad Mike just couldn't keep quiet. "With a little more training, you'd be better than your dad."

"I'm afraid my dad lost sight of the real world to his mercenary interests."

"Don't be too hard on him." Mike almost sounded sympathetic. "It's in his blood."

Ian didn't appear convinced. "So how about Colombia?"

She stared out at the land below as the red soils of the drier bushland melded into greener ranges of trees and grasses. When they closed in on Victoria Falls, the land would become jungle fed by the Zambezi River and the falls' perpetual mist. "It was a simple hostage exchange gone awry."

"You give them the money and they return the hostage?"

"In theory. Upon a visual of the hostage, money was electronically sent to a middleman. Then when the hostage was safely returned, the money was to be transferred to its destination."

Ian shook his head. "Minimal backup, I presume."

"Yep." She grimaced at the memory. "Not much different than our plan. What the guerrillas really wanted was to make a statement—a *Black Hawk Down* kind of thing—and we obliged."

"So how did you end up in the jungle?"

"They damaged the chopper on liftoff. It hit and rolled. I lost the other pilot and an American pararescue guy. The hostage and I ended up separated from the survivors during an ensuing firefight."

"Surely the surviving soldiers came after you?"

"The remaining Special Forces were mostly Colombian. Lemmon had some disagreement with the government, who pulled them out."

"Hell. So who eventually rescued you?"

The memory brought a smile to her face. "The last few days, I developed jungle fever. I was pretty delirious but will never forget the soldier who told me they were American and had come to take me home."

"This hostage you rescued wouldn't happen to be the kid you told me about earlier?"

"He was the Colombian minister of defense's son." Her heart swelled with emotion, but she shoved it away. "We evaded the guerrillas for twenty-seven days. Two hours before I was rescued, he was fishing to feed us when guerrillas came across him. I lost him. I was told they eventually killed him."

"Why did the rescue take so long? I thought the Americans didn't leave a man behind."

"Tell that to the CIA. With guys like Lemmon in charge, things get a little convoluted."

"You can't blame yourself for losing the hostage."

"The whole scenario was a lousy idea, poorly planned and set up, and impossible to execute safely. I don't blame myself for its failure. I blame myself for letting down my guard with the boy."

No doubt Ian understood how a kid trusted an adult. Failing an innocent was devastating no matter what the circumstances. That kind of pain she didn't wish on anyone.

Time to focus on more productive ground. "Describe Brugman's place to me. The likelihood is excellent you'll need Mike and me to save your butt."

Ian grinned at her challenge. "It consists of four large, circular structures connected by raised outdoor walkways and short breezeways. The central and largest structure holds three suites, a conference and living room, and Brugman's office. Somewhere in there he has his electronic toys. The river cuts across the north edge of the property from north to east. The northernmost building along the river houses a kitchen, food storage, bar and dining room and overlooks a small, rustic hot tub and pool. The eastern one along the river holds a training facility, gym, armory, and a special storage room. I suspect he has reinforced the structure since my last visit. The fourth place, to the southwest, has the staff quarters, laundry, and linen storage." Ian went on to describe various trails on the property used for security and perimeter checks and the position of the boat dock, well away from the villa so as not to block the view.

"What's the chance Kagona will already be there?"

"Good. It's better to know where your enemies are rather than have them appear unexpectedly."

"Why worry when they know exactly where we'll be, too?"

"I'm counting on it." Ian stretched out the best he could. He folded his hands under his head. "It's the only way our plan will work."

"What plan?"

"I told you, the one where you're the backup."

"This little discussion hardly constitutes a mission brief. How do we penetrate the defenses? How do we get unnoticed into the lodge or wherever Sipho is being kept? In my short and not-so-illustrious career, I've learned to recognize bad plans when I see them. Yours isn't bad—it doesn't exist."

"Hmm," he murmured, closing his eyes as though falling asleep. "My plan is simple. Get you and Sipho safely to South Africa."

Before she could climb over the seat to strangle him, Mike piped up. "His plan is a good one, missy. He's a bit unconventional compared to the old days, but so far his success rate has been high. He loves that kid. We'll get him back."

"Couldn't have said it better myself," Ian agreed with his eyes still shut.

CHAPTER SEVENTEEN

Victoria Falls, Zimbabwe

Kagona stood on a raised wooden walkway overlooking a sandy area near the Zambezi River. The walkway ended at a floating dock where Brugman berthed his shallow-water catamaran. Its three luxurious levels guaranteed special guests cruised in ultimate comfort, in contrast to other tourist booze boats frequenting the Zambezi River.

The dock vibrated under the booted footsteps of one of Kagona's men. "All secured, sir."

At Kagona's side, Brugman crossed his arms, appearing vaguely insulted. "I've dealt with Taljaard before. He won't get past my enhanced security."

"You were foolish leading him here." Frustration dogged Kagona, forced to speak through his immobilized jaw.

"It wasn't him I was worried about." Brugman sounded a bit nervous, until he looked directly at Kagona and caught the bandage securing his chin and sweeping over his head for support. Then an arrogant smirk spread across the trader's face.

Kagona squeezed his fists tight. He'd get even with Taljaard…with everyone who'd dare defy his authority.

"I'll pay your outrageous fee for the boy, Brugman, but it will cost you." The manipulative trader had pushed his limit with this latest move. Kagona had been at the hospital when he'd received news from a policeman in Bulawayo about Taljaard and the boy at the Ndebele kraal. Brugman had snatched Mukono's son before Kagona could move to capture him. "Did you figure the promised draw of capturing Taljaard would soften my ire?"

"Actually, I considered with your hasty need for Mukono's son, you'd take him and leave. I'd heard you wanted Taljaard dead. Not my thing."

"If he tears this place apart, you're going to wish you'd never interfered." Kagona shifted closer to emphasize his menacing brawn compared to Brugman. "I expect you to help me catch Taljaard. And know this—the CIO is thorough. An operative in Bulawayo is investigating how you learned the boy was a target. I will find your informant. This will be the last time a prize will be snatched from me."

"You're mistaken. I do a good deal of business with the Ndebele. I don't need a police informant when I can place a few extra dollars on a man's palm."

"You too quickly divert attention to the tribes. Your days of easy access to valuable information is done."

Brugman, well aware of Kagona's reach, had wisely protected himself and Mukono's son at this isolated African lodge. And that fact irritated Kagona.

This week had left him bested by two irritants. One had left his jaw dislocated and swollen, forcing him to wear a foolish bandage. Taljaard would give him the information he wanted and then die...slowly. The second, standing next to him, was picking his pocket for every last dime he could extort. He would be paid and left to roost, waiting for Kagona's next call to action. Both Brugman and Taljaard were bottom-feeders.

Surrounded on two sides by the Zambezi Nature Sanctuary, Brugman's lodge existed because those with power in the government allowed it. As a white man, he could operate among a different element in Zim and surrounding countries. When Kagona required something from them, he expected Brugman to supply it without the burdensome ties to government officials. The only *tourists* who visited his lodge had wealth, fame, and frequently wanted something Brugman could obtain...for a price.

This piece of shore lay sheltered behind low limbs hanging out over the water and the docked catamaran. On the sand below, three chickens were tethered separately to a metal post. Water stirred at the river's edge. The armored backs and beady eyes of crocodiles surfaced. The chickens launched into the air in desperation to flee, only to be tugged back to the ground by their restraints.

A croc big enough to devour a water buffalo emerged. Clearly familiar with this feeding ground, it crawled forward, noting the out-of-reach audience and then the meager prey. It easily snagged two chickens flapping and cackling in fright. A few crushing chomps and they disappeared into the creature's acidic stomach. A second croc moved with unexpected swiftness and snatched the last in its jaws before quickly retreating to the water.

Brugman looked pleased. "Surprising how well they move out of water. I prefer watching them tear apart an impala. One limb at a time."

Kagona had seen this cheap entertainment before. "Time is tight. Where is Mukono's son?"

"In the private dining room. I'm sweetening him up for you with ice cream and candy cake."

Kagona had given up attempting to understand Brugman. He had the cunning and balls of the strongest of men, yet avoided confrontation unless large sums of money were at stake. Even then, his weapon of choice was pampering his opponent to death.

Brugman directed one of his men to lead Kagona to the child.

In minutes, Kagona sat across a table from Mukono's boy,

who played with the frosting of his candy cake. The boy's cheeks and mouth were delicately defined, much like his mother's. The same large, round eyes that had looked to Kagona for help the night she died shone bright on the boy. He appeared mesmerized by the treat before him and hadn't acknowledged Kagona's arrival.

Interesting—a kidnapped child who showed no fear. Kagona signaled the guard to leave them alone.

The boy poked a finger into the gooey pink frosting and lifted it before his face. He slowly twisted his finger back and forth, back and forth. He licked the finger clean and started the same process again.

Ten minutes continued of Sipho dipping his finger, slowly turning it, and licking it. Kagona's patience had limits. He slapped his palm against the table.

The boy's eyes widened, then narrowed. He sat back. Pink-tipped fingers smeared sticky frosting across the tabletop on their way to grip the edge.

"Your father is safe." Kagona softened his tone, yet it came out harsh through his locked jaw. "He asked me to come get you. Bring you to him. Do you miss your father?"

Sipho Mukono stared as though his mind had twisted into knots and held his tongue hostage.

Brugman had warned Kagona to go easy, be friendly—both had proved to be useless attributes in this case. He pushed to his feet, the chair screeching across the wood floor. Sipho covered his ears and brought his feet up on his chair.

Damn child...or more precisely, freak. Brugman had softened him up with *nice*. Wasn't working. Kagona needed a change in strategy.

"Do you know who I am?"

The boy slid his arms around his legs, pulling them tight to become a ball balanced on the chair. He tucked his forehead against his knees.

"I am the man who can set your father free. Allow him to go home, so he can be with you."

Sipho showed no interest.

"You can play in your father's office. Look at the things on his desk." He leaned onto the edges of the chair and hovered over the boy. "He told me you like to play games. Memorize his papers."

Sipho lifted his head slightly and rested his chin on his knees. *Progress.*

"Do you want to play that game with me? I can be very good at it." He unfolded a paper on the table with words, names, and a series of numbers.

The boy looked away, but curiosity soon had him peering at the print. Kagona covered the bottom half with his hand. "You can't read that part yet."

The boy simply studied him, curious, perhaps, yet without expression.

"I memorize quite well, so I will go first. Your father showed me a page with names and numbers." Kagona rattled off a name and numbers. He sacrificed an account he'd memorized for this interrogation, his uncle's account in the Cayman Islands. One of those he suspected Mukono of having discovered. "Do those numbers bring anything to mind? Did I memorize them correctly?"

Sipho blinked but remained silent.

"Your turn. Give me a line from this paper. Show me you can play the game."

The boy tilted his head and stared out a window overlooking the Zambezi. A minute ticked by and then another.

Perhaps what Kagona had seen on the video was faked. What if this was simply a hoax Mukono had created? What if the boy knew nothing? The thought seemed outrageous after all Kagona's efforts, but then, desperate men did unimaginable things.

"You don't have to play. I understand how a father loves his son. We each think our son is the smartest, the best in the world. Perhaps your father bragged a bit much about your memory. Most kids can't remember things that make no sense."

No answer.

He would try something simpler. "How about we play an easier game. I will say a word. Then you say another word that goes with it." Association might drag a few unintended details from Sipho's mind.

The boy tucked his head back down.

Kagona paced several steps to cool his temper before returning to the table. Brugman, with his electronic toys, surely spied on their conversation. Kagona stood above Sipho before leaning his head close to his ear. He whispered, "I still want to play, even if you don't. My word is *Chitima*."

Birds chirped, an engine purred from a distant tour boat on the water, and damn mood music played softly in the background. A digital clock on the wall counted off thirty seconds and then another. The boy hadn't twitched.

Another three minutes passed while Kagona mentally ran through a list of techniques that worked on difficult clients. He needed a mind reader to get inside this boy's head.

Commotion sounded outside the curtain to the private room. Brugman strolled in, chin raised as if his reaching the boy first at the village made him the smarter man.

Kagona maintained his watch on the boy. "I hadn't expected you to join us."

Brugman smiled insincerely at Sipho. "I suspect you have a while left to chat with my small guest, Mr. Kagona. However, there is a *difficulty* that your men suggested I make you aware of."

The boy raised his head ever so slightly at Brugman's comment.

"Is the problem from your incompetence or caused by my men?"

"It's best we speak outside." Brugman shifted uncomfortably on his feet.

"Does it concern Taljaard?" He swore the cheek muscle near the boy's exposed jaw tightened. Had he smiled? Hope was a weakness Kagona could twist in his favor.

Brugman glanced toward Sipho. "Airing a difficulty here isn't wise."

"He should know the truth. That we've set a deadly trap for his friend." The boy remained still, as if he hadn't heard.

Brugman held back the door curtains, indicating they speak privately. "This problem is of a personal nature." He left the private dining area.

With a grunt, Kagona followed. The boy stirred. Kagona stopped in the doorway to look back.

"Tinashe Garai Tatenda Kagona." The boy spoke with his head still down on his knees. "My friend say be afraid."

The boy rattled off a number sequence, ending with, "…five oh six two two Cayman Islands." Kagona only remembered the last four of that account, but he had no doubt the child had repeated it correctly. And then he called off twenty-one more numbers from a Swiss account.

Kagona reeled at the implications. Had any of the numbers been handed over to his opposition or an unfriendly country? Had the Americans already locked down his accounts? Could he move his money out in time? Did Mukono or the boy know the truth behind Chitima?

With his pulse pounding, he let the curtains fall in place and went to find Brugman.

The man grinned when Kagona caught up. "I believe the boy's price has just gone up. He gave you the proof you wanted. There is no need now to wait for Taljaard."

"Are you afraid of what one man will do to your compound?"

"After much thought, I realize this is an exclusive lodge, not a war zone. I've had a run-in with him before. You might say it was costly."

"There are no other guests to witness events, and this time with my men present, Taljaard's capture will proceed swiftly...unless you fear he's after you personally."

Brugman swallowed hard. "He doesn't scare me. Once you catch him, please take your men and leave. I have a reputation to maintain. You should've taken Taljaard at the Ndebele village. He was under the influence of the *sangoma*'s sleeping drug."

"A point you didn't mention in your message about obtaining Mukono's child."

"What did you expect? You could've come to me in the first place about the hunt for the boy, instead of forcing me to glean your interest in him through costly channels."

Brugman whined too much. The boy's tongue had loosened. Now was the time to dig for more. "What is your crisis that couldn't wait until after the boy's interrogation?"

"I've detected an unauthorized radio transmission from my lodge."

Kagona snorted. "Taljaard is on his way. Have you found the source?"

"We made a search of the premises and discovered a radio hidden in an employee's locked trunk. A shame, really. He was my best acquisitions man."

"Your guard is down if one of your men is working against you. Where is he?"

"In my office. He's not being cooperative. Claims he's innocent. I'll have him sent back to his village. He's no longer welcome here."

"I've a better idea to encourage both him and the boy to talk. Soon they will be singing music I want to hear."

"I don't handle things the way you do, Kagona. No killing." Brugman looked like a man fearful of losing control. He was.

"*I* don't plan on killing anyone. But I want the boy to tell me everything in his head. And surely you want the same from your man." Kagona hit a button on the walkie-talkie attached to his belt. Footsteps pounded toward their position. His man Chaipa appeared.

"Bring the boy. He's coming with us." Kagona pointed at Brugman. "Get your man. We'll see how long it takes him to talk once he's staked out as meal for your pets."

Brugman blocked his way. "That's not how I operate."

"My men may be few, but they have no qualms about taking on yours. They, like me, are skilled fighters. Do we work together and make your problem go away, or do you want to make a bloody mess of your precious lodge?"

"Be careful with your grand scheme." Brugman stepped aside. "It might backfire. Taljaard is popular."

"Your man can save himself by telling us who he was contacting with the radio. It's about time your cadre learns what it takes to make a good living." Kagona brushed past Brugman. "You needn't worry about cleanup. If he doesn't talk, there won't be a body left behind."

Kagona strode out and down the path toward the dock along the Zambezi—the place teeming with dangerous life, like crocs and hippos…and humans.

Rich shades of green covered the basalt rocks where Victoria Falls, one of the Seven Natural Wonders of the World, plunged into a gorge carved by the Zambezi River. Mad Mike aimed the chopper at an ethereal spray rising a good thousand feet above the falls. On a normal sightseeing tour, Joni guessed, now was when he opened the tiny windows so his passengers could better hear the roar, the result leaving them delightfully wet.

Ian awoke from his lengthy snooze. He reached a steady hand past her seatback and reassuringly touched her shoulder. "First view of the falls?"

"I've seen it from forty thousand, but this is my first view up close." She swallowed hard, thinking about Sipho hidden somewhere along the broad ribbon of water above the falls.

The view eased her tension about the unknowns ahead. From behind, Ian gently wrapped his fingers around her headset. He coaxed it off her ears and down to her neck with the same finesse a man might use to unhook a bra or slip a shirt over a shoulder. The removal of such a common object in her daily life left her feeling oddly vulnerable.

"Hey"—he spoke directly into her ear—"if there is one thing I know, it's the good, the bad, and the ugly about the people that populate my country. I'll get Sipho to you. Then it's your turn to get him out of the country in one piece."

His closeness felt both reassuring and disconcerting. It had taken her two years to reach some sort of acceptance within her own mind that she had done everything possible to save the Colombian defense minister's son. The kid had stuck with her for weeks, sharing his fears, his dreams, his hopes of a future. Then he had none. Her eyes watered.

She was on edge again. "I might feel better if you gave me a few more details."

"Sure, but you won't like what I tell you."

"Humor me."

"It's not the least bit funny. Kagona has a personal score to settle with me. Since his men discovered you in Bulawayo at the park, he'll figure I'm not far away. He'll assume any rescue of Sipho from prison in Harare would be near impossible for me, so he'll leave him in a place where he believes I'll make an attempted rescue. Brugman has provided the perfect place."

"So you're expecting a trap."

"Yes, and it means there's a decent chance you or Mike will be caught. That's why we're splitting up. Sipho is my first priority. Getting you guys out, if necessary, will be secondary.

Therefore, you'll need to use your own resourcefulness and not get captured or killed."

"You're right, I don't like it, but I understand. It also suggests a dozen other questions."

"None of which I'll answer, simply because I fully expect Kagona to do his best to coerce the details from you, should you be captured."

"What's a little torture?" She turned back to the terrain below, noting Mike had skirted the mist. The sun hovered just above the horizon. They'd be taking on Brugman's villa in the dark. "I'll need a weapon."

"Is a nine-millimeter Browning good enough? It could get you killed."

"Would you go in naked?"

"This time, no. But I won't get caught."

"Neither will I."

"If Kagona is there ahead of us, he'll appeal to your weaknesses. Expect it and be prepared."

"And what weakness might that be?"

"You're a woman and American."

"He doesn't know that. Nor does that make me soft or stupid."

"He's no fool, either. Enough people have had contact with you to know you speak English. Exactly which country you're from is probably not important. But where Brugman prefers to avoid bloodshed in his endeavors, Kagona has no qualms about it. Brugman thinks about the bottom line, but Kagona, in his twisted mind, thinks about the glory of his country *and* his pocketbook, and will do anything to protect them."

Mike circled a prominent Victorian structure as though intimate with its landing pad. He did a quick check for obstacles on the pad and noted the wind sock's direction. A broad, rolling lawn would make for an easy touchdown.

She slipped her headset back on.

"The Victoria Falls Hotel," Ian said. "Built in 1914 for the first tourists. If you have bucks, you stay here. If you don't, there are other, less grand places along the river. As you can see, it's a short hike to the falls from here."

The sheer volume of tumbling water held her enthralled. "Pretty amazing sight from the air."

"You should see it at night." His husky voice told of vivid memories. "The moon and mist make a lunar rainbow over the falls." Ian tapped her shoulder. "Come back sometime and I'll make sure you see it."

"That's a date," Mike cut in. He prepared for landing. The skids had barely touched down when he idled the throttle and flipped off switches.

The rotor slowed to a comforting *whoosh, whoosh.* Outside, a multiseat cart from the hotel drove in their direction. "It's going to seem a bit strange with bedraggled passengers arriving at such a posh hotel."

"Naw, that's my mate Stanley." Mad Mike popped open his door. "He'll help us out."

The cart swung up after the rotors stopped. Mike unloaded the three packs. Ian slid the heavy jerricans to the door and then helped Mike hoist them to the ground. Time to set up for their quick getaway.

Joni hooked up grounding wires, and they refueled the chopper. Once done, Joni wiped spilled fuel off her hands and Ian reloaded the cans. Mike vented a window in the Jet Ranger before chatting with his buddy. Although subtle, money passed hands. He also shot a concerned look toward Ian.

Ian and Mike took a moment to converse out of earshot before they all loaded into the cart. Instead of delivery to the front door or registration, their driver dropped them at the far edge of a parking area.

"Hold tight, kids, transportation's on its way," Mike said before walking away. Minutes later he rattled up in a battered sedan built before Joni was born. "Hop in, mates. The ride is short."

"Thank God for that." What that car didn't belch in fumes, it vented inside for the pleasure of its occupants. By the time Mike pulled up at an old place along the river, Joni's head pounded from inhaling carbon monoxide.

Not waiting for an order to abandon ship, she launched out of the vehicle and gasped in humid, fresh air. "I can see it now. We succeed in the mission only to succumb to fumes during our getaway."

A tour boat floated alongside a dock made of bright yellow drum barrels connected by metal planking. The entire dock seemed mobile, to keep up with changing seasons when the river flow drastically altered.

An expressionless Ian watched Mike speak with a man who managed this makeshift marina. Mike pointed at two small boats with upturned hulls parked under a purple jacaranda tree. The manager shook his head. After another minute of continual head shaking and arm flailing between the men, Mike pulled a wad of cash from his pocket and counted it into the man's hand. Faster than water over a fall, the cash disappeared.

Mike motioned them to bring the packs. Ian slid one, bulky but not too weighty, over her shoulders, giving her no chance to inspect its contents.

Joni pointed back at her pack. "Explosives?"

"Nope. Mike has those. Try to make sure he doesn't blow himself up." Imagining Mad Mike loaded with explosives did nothing to quash the nerves starting to surface.

Ian slapped the back of her pack in a send-off much the way football players did to their teammates' backsides upon taking the field. Did that mean he had accepted her as a teammate—or was she about to get creamed by the defense?

CHAPTER EIGHTEEN

Kagona's men clamped the iron about the traitor's leg. Brugman argued with his man. The guilty bastard simply spread his hands, shook his head, and pleaded innocent.

Mukono's boy dropped to the dock and retreated into a ball again. His strange behaviors had become annoying.

Brugman looked up at Kagona from beside the man who supposedly had been his top performer. "This isn't necessary. Frightening the child won't help your cause. I haven't had a chance to thoroughly investigate this man's claims of innocence."

"Ring the dinner bell."

One of Kagona's men stood on the dock by the catamaran and slapped the water with the flat end of a paddle. He continued while the man with his ankle chained to the feeding pole argued his case. The slapping took on an eerie rhythm with the victim's babbling voice.

Brugman hastily climbed to safety on the raised walkway. The water became shadowy as the sun dipped low. Only minutes until the sun set.

A light came on at the dock. One of Brugman's men stood on the deck of the boat, manning a spotlight and waiting for an order to turn it on.

Water splashed, and a soft swoosh sounded—a heavy object being dragged on the sand.

Kagona leaned down to his little friend hiding at his feet. "You can stop this if you tell me about the accounts. All the numbers you saw."

The boy simply covered his ears and buried his head.

Three crocs emerged from the Zambezi, creeping onto the sand. The men remained on high ground, well aware of their swiftness. The tethered man grabbed the pole, putting it between him and the crocs. His pleas became frantic, mesmerizing. Kagona understood the attraction of animal events to the ancient Romans.

The man had gone from yelling his innocence to self-preservation. His attempt to climb the pole brought a croc roaring forward. It missed latching onto the man, but in its leap its foot had caught the chain and shortened the length by a good foot.

The spotlight clicked on and shone on the man. A second croc lunged from the other side and clamped tight on the man's leg, bringing him to the sand. Kagona's heart raced at the screams. The croc dragged the man until the chain stopped its progress. One croc held him tight while another latched onto an arm and twisted, tearing it off.

The boy, curled at his feet with his ears covered and eyes shut, started singing.

Power surged through Kagona. "Come and get us, Taljaard," he shouted. "We're waiting for you."

The closer Joni, Ian, and Mike came to their water transportation, the more she feared their assault would end at the bottom of the river before they ever made it to Brugman's lodge. The upturned dull-green hull had numerous fiberglass patches and looked more like a quilting project gone bad than the sleek aluminum boat that sat next to it. Mike's choice, although ugly, made sense. It didn't take much moonlight for something shiny to gleam on the river.

Something Ian carried in his pack was heavy, but Joni knew better than to ask. Mike and the manager flipped the boat upright and set about mounting a motor, while Ian dropped the packs aboard. She wandered to the water's edge, where they would put it in.

"Don't stand too close to the water, missy. Crocs. They hunt at night. Spring out hard and fast. You'll never see 'em coming."

Whether Mad Mike droned out a tall tale or told the truth, Joni moved back from the water. Everyone in Africa knew crocs killed a good number of fishermen and villagers every year. She had learned alligators left boats alone from canoe trips in the Everglades, but had no idea whether the same held true for crocodiles.

Mike and the manager grabbed the gunwale on one side, ready to lift the boat into the water. At their position, the mighty Zambezi stretched across a wide, braided riverbed and moved slow enough for easy navigation. A mere quarter kilometer downstream, however, the Zambezi River dropped over Victoria Falls and roared deep and rapid through narrow gorges.

She and Ian picked up the other side and the four carried it to the water. Ian climbed aboard and positioned himself at the bow. She boarded next, keeping low and settling on a central bench. Mike shoved the boat into deeper water before leaping on and sitting near the small outboard motor.

Mike steered them upstream. A splash nearby reminded her to tuck her arms in close and sent her scouring the river for long, dark shapes. As twilight faded to orange, various birds flew toward small islands centered in the river. Some had trees for a safe roost while others were simply bare basalt rock.

"How far up river to Brugman's?" she asked Ian over the engine's buzz.

"About twenty minutes without interruption."

What kind of interruptions did he expect?

Mike navigated relatively close to shore. At least if their mighty piece of transportation sprang a leak, they were close enough to land to beat out the crocs. Maybe.

Ian fished out a bundle she could hardly discern in the dimming light and started passing items back. Joni took a black ski cap and a lightweight pullover and pants before passing the extras to Mike. As she turned back, Ian, now a shadowy silhouette against the river, shrugged off his buttoned khaki shirt.

Defined deltoids wrapped around his shoulders and drew her eyes to strong upper arms. It took dedicated training to maintain that sort of physique. The thought made her question, once again, what had been left out of his dossier.

Ian faced her. "Suit up, we're getting close."

"I'm afraid I can't offer the same caliber of display." She yanked a shirt down over her head. The extra bulk might even fool someone into thinking she had more than 120 pounds under the clothes.

Ian slid his legs into dark pants and worked them up over his shorts.

"I'm disappointed," she said, "I expected to see more."

He chuckled and strapped the pant legs closed around his boots.

"Enough flirting, kids," Mad Mike chimed in as he swung the boat even closer to shore. "Lights ahead."

Sounds of music, laughter, and voices came from up ahead. She tugged on her dark pants without rocking the boat.

"Booze barge dead ahead and coming our way." Mike tugged down his cap. "Better if they don't get a good look at us."

Joni followed Mike's example, donning the hat and tucking up her hair. Darkness had almost enveloped the river.

Ian opened a chewing tobacco-size container and dragged his fingers through the contents. With a few quick swipes, he spread dark camo paint across the high points on his forehead and face. He handed the tin back to her. "Pretty up and then hand the stuff back to Mike."

She scooped up a hunk of paint with two fingers and copied Ian's handiwork. Using a color stick, he added another streak to the low points of his face but didn't bother to hand the stick back.

The tour boat puttered past. A minute later their craft rocked in the bigger boat's wake. Ian had grown quiet, continuing to deck himself out with paraphernalia he retrieved from the pack. Night-vision goggles hung around his neck, and a thick vest and belt had various odd-shaped objects attached to or stuffed into them.

Behind her Mike rustled around in his pack. "I suspect you know how to use this?" He handed her a pistol secured in a nylon holster.

"I'm a tad rusty." The weapon felt cool in her hand. She checked to make sure a round was chambered, then slipped the adjustable belt around her waist and snugged it tight.

Mike had one more weapon to add to her arsenal—a knife in a Velcro wrap. She fastened it to her ankle. The knife did little to boost her self-confidence at raiding a fortified compound. Surely Ian had something more substantial. She reached for the pack she'd carried onto the boat.

Ian caught her wrist. "Better if you don't know everything." His firm grip exuded control and confidence.

"Are you sure it isn't wise to give me at least some hint?"

"I'd like to. I just don't trust Kagona. What's in there says way too much about my plans." Instead of releasing her, he stroked a finger across her palm. "Keep yourself safe. You may be our key out of the lodge complex."

He might as well have swiped his finger somewhere much more personal. One moment she hated him for his cavalier attitude that he had everything under control, and the next she wanted nothing more than a little downtime with him. Time that had her body pressed against the solid form she'd watched disappear under that pullover, where worldly problems didn't exist.

She tugged her hand away and looked out across the river. It had become a black velvet ribbon. A few stars flickered hazily, affected by the mist created by the falls. Hopefully the moon would make an appearance soon. Its quarter last night had served to illuminate their road trip to the village.

Ian handed back a wide-band watch. "It's a GPS unit. Used one like this before?"

She pressed one of the buttons, and a red digital readout appeared. "Have an old model at home. Not meant for stealthy night use like this one, though."

Mad Mike chuckled. "The best feature on that is the airport search. Gives me a list of the closest airports in case I get lost."

Ian snickered. "I've never seen you lost."

"See, I told you that's all I need."

How Mike maintained a sense of humor eluded her. Perhaps that came after surviving a dozen missions with lousy odds. If Mike had been flying for forty years, he'd done most of it without global positioning satellites and a large amount by the seat of his pants.

Mike donned NVGs like Ian's to aid in steering in the dark. At least someone could see where they were going. Hopefully the men planned to offer her a pair. She hated to think she'd be relegated to stumbling after Mike.

She keyed the GPS to leap ahead to their destination. The large watch face shifted a map view along the river. According to Ian's checkpoints, they hadn't far to go. "When did you have time to program this? I didn't see you working on it in the chopper."

"I had the route preprogrammed from an earlier visit."

"From that intimate adventure you and Mike mentioned earlier?"

"The very same."

"How often are you involved in these dangerous adventures of yours?"

"When I was running my ranch as part of a conservancy, Brugman and I were frequently at odds. But I didn't undertake anything this extreme with him until recent years. The man will do anything for money, including illegal hunts, and I'll do anything to stop him."

Ian touched her hand again, hopefully as much from personal desire as the need to maintain her attention. "That will get you back to town if we get separated." He positioned the goggles on his face, then signaled everyone to silence.

Mike cut back on the engine power. The boat barely made progress against the current. More stars had appeared in the royal blanket above, and their meager light reflected off two glowing eyes in the water dead ahead. She signaled Mike and pointed them out.

He changed course slightly, moving the boat closer to the shore. Deep, guttural grumbles and burps carried across the water. He leaned forward and whispered, "Hippos."

Great. A chance encounter with hippos and the three of them would be permanent river dwellers. The glowing croc eyes dipped into the river. She estimated twenty feet to its last location. Any time now they'd be passing over it.

Something thumped hard against the bottom of the boat. Water trickled under her boots.

"Shit," floated on the night air back to her ears. Ian lifted the pack under her seat and set it on her bench seat to keep dry. "One minute till touchdown." Noting no panic from the men, she assumed they'd survive—but would the boat make the trip back?

Mike cut the motor, and they coasted to shore. Ian jumped out as the boat grounded on a steep, sandy slope. She snatched a pack and followed. Mike landed right behind her. She grabbed the boat and helped to heft it to dry ground.

"We'll hide it in the brush up the slope," Ian whispered, but then he halted. "Drop the boat."

She heard him move off along the crest of the bank and then return.

"Brugman knows this is a good place to clear the river. He has laser trip wires set up between two trees. To avoid the crocs, he set them waist-high. We'll slide the boat under. Joni, go first."

She took off her pack and dropped it where Ian stood. On hands and knees, she pushed it past the point he indicated as hot and crawled after it. The men shoved the boat under the laser. In less than a minute, they had the boat hidden in the brush.

Her heart pounded with exertion. Ian picked up her pack and slung it over his shoulder with his. "I need this more than you." She sensed the urgency in him.

Green light flickered as he removed his NVGs. In a surprise move, he cupped her chin in one hand, holding it as though it gave him strength. Torn between feeling uncomfortable with his touch and longing for it, Joni reached for his hand.

Before she could connect, he hung a plastic earpiece over her ear with a short extension to her jaw. By placing her hand over his, she helped him set the listening pod in her ear. He then slipped a cord around her neck with the rest of the unit.

"It's set for push to talk." He placed her fingers on the button.

She stuffed the unit under her shirt to keep it from swinging free.

"Wish me luck." Ian slipped on his goggles and walked into the night. He moved so silently, she had no idea in what direction he had gone.

Mike stepped up to her side. "Did he say 'wish me luck'? Never heard him say that before an operation." Concern actually sounded in his voice. "You sure you're good to go on this?" he said to Ian over the comm.

"Never better."

Before anyone could reply, Ian came back quietly with,

"Company at the dock guarding the boat. Two. Brugman's. Two rifles, NVGs, a few handguns. Expecting us along the river."

"We'll be careful," Mike answered. He handed Joni a pair of NVGs. "These are old ones, but they'll have to do. I'd give you mine, but they cost me my last dollar and I don't even let Ian touch 'em."

Joni settled them on, unsure how to adjust the focus for ground work. When flying she set them for far distance. Fortunately, whoever last wore them had decent vision. She'd make do.

She tapped Mike on the shoulder and pointed back to their boat. "What about the leak?"

"Just a small hole. I can plug it."

"Did we hit the croc?"

"Naw, just a root from a water berry tree. It was overhanging the river. Didn't you see it?"

No, and she wasn't born yesterday. Mike didn't want her worried about crocs on the race back downstream.

"What now?" she asked.

"We head to the dock."

"Isn't that where those men are?"

"Yep, and where Brugman keeps his boat, which can track us down on the river. We need to make sure it doesn't follow."

"If it's faster, why don't we borrow it?"

"Mr. Brugman likes to hang onto his toys. He probably has a remote engine kill switch installed. It'd take me too long to find it. I think we'd have a better chance in our own."

"With a hole in the bottom?"

"You're getting to be right cocky, missy."

"It's the company."

A partial moon had risen above the tree line, creating shadowy outlines when she looked skyward. Mike led her into denser brush, as though believing slow and silent a more useful plan than fast and exposed.

197

Glimpses of the dock appeared behind a gathering of tall palms. The double row of floating drums with metal planking laid across them didn't jut out into the river but rather paralleled the shore. A sleek, powerful catamaran longer than the dock rocked gently against its mooring. Brugman indeed had the best of everything. This beauty wouldn't simply overtake the piece of junk they'd arrived in, it would mow them down and feel nothing more than a slight bump.

Under a single light positioned high on a wood pole attached to the dock, a well-armed man paced. He was dressed in jungle-green fatigues with a bulletproof vest and thick belt to which a holster attached. Whatever he packed in the holster looked as threatening as the submachine gun he cradled in his arm. Ian had said two men, but she only saw one, and this one was enough to raise doubts about the simplicity of the mission.

Mike signaled her to stop. She knelt and held her position. Night bird calls echoed in the distance, nearly obscuring the steady sound of water hitting the dock barrels.

Something large rustled a mere twenty feet away. Her NVGs highlighted the distinct shape of a crouching human. He must be the first of Brugman's men Ian had spotted. The man waited and through his own NVGS watched the clearing and shore along the river.

Mike slipped off his pack and left it with her. He pointed to himself and then to the man ahead and the one at the dock. He had to take them out before he could disable Brugman's boat. She signaled him okay and stayed in place while he disappeared silently into the night.

Joni shifted to better see the dock and to use a tree trunk to shield her in case Brugman's man looked back. This time her view of the dock revealed a disheartening setting.

A metal pole stood ten feet from the dock, with a length of chain hooked to the lower portion. The chain led off into disturbed sands and had a shackle attached to the end. Grisly remnants left in the iron looked suspiciously like a leg. Human?

Shit.

The disturbed sands and proximity to the river ran her imagination wild. Brugman, the guy who supposedly believed in money and not killing, had fed someone to the crocs.

Disbelief sickened her stomach. Had Ian had made another error in judgment? Was his perfect plan really so great?

Brugman expected Ian to rescue Sipho. Surely they anticipated that if Ian arrived by boat, he might try to steal or disable Brugman's. This setup made a perfect trap. As such, the men must be in communication with one another as well as with Brugman. They would know within minutes—hell, seconds—when Mike took out the guards. She and Mike had to move fast.

Joni scanned for the man nearby. Nothing. She didn't dare try to raise Mike, as it might reveal his position. The last thing she wanted to do was stand on Brugman's doorstep before it was necessary and yell, "We're here."

Again she was relegated to waiting and hated every second. Action had much greater appeal.

Ian's comment about Kagona using people's strengths against them ran through her head. Did Brugman also follow the same principle? If so, was there more to this trap than they could see?

The guard at the dock paced close to the boat. The sleek hull gleamed white. Chrome accents on the three decks reflected the dock light. A plank ran through an opening in the shiny safety rail on the lower deck and welcomed passengers aboard.

The guard swatted at his neck. The next moment he looked at his palm before keeling over.

"Okay, that's the two of them," Mike said over the comm. "I'm ready for the pack. Take it slow. No saying there are not more men out here. There's cover till about ten feet aft of the boat. Meet you at the water's edge."

"What about the crocs?"

"Hurry, missy, or I'll be dinner."

Mike was too gristly for the crocs. She, on the other hand, would make a tasty snack. These scaled creatures gave her nightmares.

The pack weighed heavy on her shoulders, but she took her time and continued a sweep of the area as she moved. The last ten feet left her exposed, but if Mike had done it successfully, so could she. Joni made the dash and breathed heavily as she met Mike, sheltered by the boat at the water's edge. He lifted the pack from her shoulders. "I'll need a lookout while I work," he said. "It shouldn't take long."

"Wait. Did you see that chain on the shore with human remains?"

"Best to put it out of your mind. We already know these are not nice people." He waded into the water above his knees to reach a low platform on the stern of the boat. She scurried after him, not wanting to present an easy target for a hungry predator.

Mike boosted her aboard the back of the boat and she in turn offered him a hand up.

"Missy, find a sheltered spot on deck where you can see the dock and beyond. Don't touch anything and let me know if you see anyone coming. I'm heading to the bridge and then below deck. I won't be long."

Mike set to work while she scanned the night. Occasional sounds from the water reminded her about the glint of croc eyes poking above the river on their boat trip here. No eyes glimmered on the water.

Hardly any time passed until Mike appeared as silently as he'd left. "Done. Let's get the hell outta here. These men had a check-in time and it's likely past." He slid down the back of the boat into shallow water, the way they had come.

More unidentifiable sounds came from the murky river. She opted to hurdle over the side near the stern, which stuck out past the dock. With a little momentum, she'd land on the shore instead of in the shallow water. No reason to make it easy for the crocs.

Joni took quick steps for a running leap. Her hand went out for a boost from the shiny boat railing.

"No, missy," Mike yelled through her comm.

Her feet left the boat deck and hand contacted the metal railing. A powerful voltage flowed through her body so violently it blasted her through the air and left her airborne, with no control over the landing.

Her limbs refused to break her fall as she headed toward the light pole attached to the floating dock. Her shoulder impacted the metal. Her body whipped around the pole like a rag doll and landed supine on the dock planking.

She fought to draw a breath as a giant force stomped on her heart. Frozen in place and staring upward at the light, a breath finally came. The acerbic smell of singed hair and the odd convulsing of her muscles created a surreal haze.

The light above flickered on and off. As though to comfort her from pain, it finally blinked out. She heard boots pounding in her direction before blackness devoured the bright sparks dancing before her eyes.

CHAPTER NINETEEN

Shit, shit, shit. Ian duct-taped a flash bang into place and hooked fishing wire to the ring on the pin. He stretched the line across a walkway.

"Can't tell if she's alive, but the troops have arrived." Mike sounded shaken over the comm. "That was a hell of an electric shock Brugman rigged to the railing."

The same wave of helplessness as when Sipho had disappeared washed over Ian. But he had his priority—Sipho. They'd all agreed. So why in the hell did it hurt so much?

At least he counted on one thing—Brugman loved to gloat. The bastard's self-aggrandizing would buy Joni time until they could get her out, if she still lived. Brugman never passed up the chance to savor every victory, especially after Ian had bested him last year. The wild card was Kagona.

"She'll be okay." Bloody hell, did he really believe that? "We've no choice but to leave her for now. She's on her own until we reach the objective. Is comm jeopardized?"

"Possibly. My guess is her unit's fried. Be careful what you say."

Ian imagined Joni lying helpless and vulnerable on the dock as Mike described. Damn it, Mike should have warned her sooner. He knew better, and if she had any training at all, *she* should've known better.

That trap had been meant for him. They knew he'd come for Sipho. Brugman treated life as one big game. The more he scored off people, the more money he had to show for his efforts. The more money, the more toys and gadgets he acquired to ply his perverted tactics on the helpless or desperate. To him, the people who suffered were just characters to manipulate on the way to winning. This psycho needed a reality check, one that included a taste of his own voltage.

Ian checked the time. Priorities first. He moved toward the next location to place a small but effective charge. Joni's capture wouldn't change the timing, just the workload. So why then had his heart accelerated to the point of pounding in his chest?

"Missy's been collected," came over the comm. "Three of Brugman's men."

"Alive?" Hope pushed away guilt building from his own failure to keep Sipho secured, which had set this entire series of events in motion.

"No movement, but Brugman wouldn't want to see a dead body."

"He doesn't need her to set a trap for me. Sipho is his ace in the hole. He'll want to brag." Joni was simply water on a dry savanna to Brugman. He'd enjoy her company, get a few bucks for her, and qualify the event as another victory.

If she survived with her wits intact, she'd be pissed and make Brugman's gaming unpleasant. Brugman couldn't command her with a controller or a joystick, and she certainly wouldn't go in any direction he desired. In the last day, she'd been hustled, lied to, and tricked beyond any person's endurance. Everyone had a limit. He'd love to be a cockroach on that wall.

If, if, if. He had to believe she was okay.

Kagona, though, was another matter. He thrived on torture and death, all in the name of Zimbabwe. Citizens like Sipho or his father or even Ian and his family didn't count. They had a different vision for the country, and Kagona believed there could be only one—his, or at least the one of his political bosses. And theirs included the right to line their own pockets. What Sipho held in his head could put a damper on that so-called right.

Sipho had memorized more than data on illegal skimming, blood diamond proceeds, or politicians' secret accounts. He'd seen numbers related to an unknown line item—a fund that had been tagged *Chitima*. Mukono could determine no official designation or reason for the funds. Nothing that showed the money as coming from government taxes or income to the state. The money's origin raised major questions, ones that Mukono had asked Ian's contacts to trace. Nobody liked the answer that came back.

The day of the accident, Ian had come to Harare at Mukono's urgent request. His friend had learned something new but didn't dare speak over the phones or internet. With so many staff and family around, Mukono wanted to wait for late that evening to relate his findings.

Mukono and his wife never returned from dinner. Whatever information he'd collected had been destroyed in the safe, unless it was locked in Sipho's head.

The Americans feared Kagona and his cronies had something planned that could change the balance of power in Africa and the Middle East, and its success necessitated complete secrecy. Kagona had to discover if details of his plan had reached beyond the borders of Zimbabwe, whether he dug it from Sipho's head or used him to make his father talk.

Kagona had arrived at the lodge aboard an Alouette helicopter. In all likelihood he was speaking with Sipho at this very moment. Ian had noticed one of Kagona's men and suspected more roamed the grounds or stood as bodyguards. This time his men weren't local police with no training and little reason to fight. They were trained CIO operatives, handpicked by Kagona.

Kagona's fast transportation to the lodge had one redeeming value—space was limited. Kagona planned to take Sipho and Ian back to Harare, so he'd likely brought fewer men with him. Ian figured about four, including his pilot, unless they added one extra man and expected Sipho to sit on someone's lap. Kagona would count on Brugman to cover the needed manpower.

Brugman had about ten men and Mike had disabled two. That left eight or so. Still a good total to worry about, but Ian found confidence with the decreasing number. Brugman's men worked for money and preferred to keep clean hands. Kagona and his men sought power, a much scarier goal. Ian would rather challenge Brugman on his home turf than Kagona.

"All set here," Mike broke into his thoughts. "Go with plan B now that missy's occupied?"

"Hell, yes. She's bright enough to survive till we can help. I hope," he added under his breath.

Ian scouted the place where his latest intel reported Kagona had secured Sipho. A visual on him would be impossible until the last moment. Ian had to count on his insider. The snatch had to be quick, quiet, and unexpected.

An odd shaking jolted Joni awake. She popped open her eyes to see the world spinning around her. Her stomach tried to hurl. She scrunched her eyes shut and didn't dare move. How much damage had her body incurred?

Pure willpower settled her insides down. This time she cracked open her lids slowly. Several feet below her head, a dirt path moved in jerky unison to her body. Her chin bounced against the back of someone's damp shirt. A foul stench of body odor agitated her unstable stomach.

Her senses regrouped and put a picture together. She was being carried sideways. A man in a red cotton T-shirt had an arm looped about her torso. Another had her thighs and knees tucked firmly under an arm.

Her scrambled brain should have seized in fear, shooting panic to every cell. Instead, it assessed her condition, waiting to send the alarm until sure the body could actually respond. A result of Brugman's excessive shock therapy?

The men left the raised walkway and stepped down to a flagstone path that cut through a lighted garden courtyard. Fragrant plants overwhelmed the stench of sweat. They passed flowering hibiscus, bougainvillea, and plants she didn't recognize from her skewed view. Such service. Personal escorts for a tour of Brugman's paradise.

An open stretch of lawn appeared as they stepped back up on a walkway leading to the main building. Beyond it to the north sat a familiar-looking Alouette. Kagona was here, and she hadn't had the chance to disable his rotors. A profound sadness enveloped her. She had failed.

The enemy had exploited her weakness and captured their prey. Now she'd have to face Brugman and Kagona. Lesson learned…if she had a future to ever use the knowledge.

Their arrival had been too easy. Brugman wanted them to walk into his lair. She'd obliged and stepped right into his trap. Zapped into submission.

Focus. Feeling sorry for herself wouldn't help her situation or Sipho's. Ian had said to survive until he could get her out. That she could handle.

Her body bearers shifted and revealed an additional gun-toting thug in their parade. No Ian or Mike. Hope gave her something to work with in the interrogations that most likely waited.

Joni tried to lick her lips, only to find her tongue completely dry and glued to the roof of her mouth. Raw pain shot from the hand that had grabbed the railing, while the other hand refused to do more than barely flex tingling fingers.

This was not good.

The door on a massive thatched building opened to a blast of cool, dry air. With the African village look on the outside and Mr. Brugman a hunter of money, she'd expected the inside to be some kind of animal trophy place. She guessed wrong.

Her bearers hauled her through a cavernous room that could have served as a five-star hotel's lobby. Brugman gave new meaning to classic safari decor. Modern chrome, glass, and rock mixed with wood and animal-skin furniture. Downward-facing spots highlighted African paintings, woven baskets, and hand-carved and molded art objects. Stained wood beams supported the biggest thatch roof she'd ever seen.

Automatic doors slid open on a metal track. The entourage left the big cave for a comfy room about the size of her apartment, with space for a conference table at least a dozen people could easily gather around, a complete office suite of glass shelving, a credenza, and desk, and three stuffed croc-skin chairs. Her hopes for one of the comfortable seats ended when the men dumped her right side up in a straight-backed wooden chair.

They pulled her hands behind the chair back and secured them with nylon rope. Guess Brugman's supply of handcuffs or zip ties was running short. That the rope hurt said her body was still thriving. She tried moving her legs, and they only marginally cooperated.

Joni stared at the back of a tall leather office chair behind the desk. It moved ever so slightly. Whether Brugman or Kagona sat there, she couldn't tell, but she suspected someone would spin around in a dramatic moment meant to intimidate. Men were so easy to read.

Behind her someone whistled. The desk chair swung around, and a black face fringed in white stared back at her. The monkey hurled its speckled gray body across the desk and scrambled up a chair. In an easy leap, it launched into the arms of a scrawny, pale man standing in the doorway.

"Ah, Bveni, I see we have caught the first fish in our net." The monkey grabbed a piece of fruit the man held in his hand and munched down. The speaker, whom she assumed to be Brugman, handed the monkey off to one of his men and moved in for a closer inspection.

"My, my, and a pretty fish, too, under that painted face. I'm glad to see you are in one piece. I hadn't planned to kill you. I used DC current. It tends to throw people away from the shocking object rather than glue them to it. I did set the current for someone a bit bigger, though. Like Taljaard. Kagona is paying top dollar for him and the boy. You, well, he wants you alive long enough to meet."

Thrilled, she attempted to say, but with her dry mouth only a mumble escaped.

Brugman sized her up from head to toe.

She returned the favor. Fortyish, he was garishly dressed in casual slacks and a pressed short-sleeve shirt. His Rolex glimmered, and a dazzling gold-and-diamond ring matched the buckle on his exotic, thin leather belt. Pale yellow socks coordinated with his shirt, but his expensive leather loafers rather clashed with the combat boots worn by his fatigue-clad henchmen. Must be casual Friday for the troops.

He wasn't at all the brooding bully she'd imagined from Ian's description. Compared to the others in the room, Brugman was short, and skinny enough a stiff breeze could easily blow him over the falls. Perhaps that explained his egocentric necessity to divulge his brilliant role in her capture. He was compensating.

A rather confused look melded onto his ruddy, freckled complexion. "You're not at all what I expected," he said. "Kagona claimed you outran a helicopter. Took out two of his men."

Right, that's me. Indestructible. The back of her head and shoulder started to throb. A vague memory rose of hitting something in her sizzling fall.

With concerted effort, she loosened her tongue from the roof of her mouth. "And you believed him." She started to shake her head, but then thought better of causing any more brain damage. "Your mistake."

"No mistake at all. One of my men is recovering at the dock. The other is missing."

"Perhaps he strayed too close to the water. Crocs can be dangerous."

"I seriously doubt you alone caused them trouble. I suspect Taljaard is out there somewhere, isn't he?"

"I think you underestimate me."

Brugman waved an arm at his man. "Oh, no, no. I never underestimate a woman's ability...to lie." The way he flitted around, she harbored some doubts as to his degree of masculinity. He snapped his fingers, and his flunky dumped several objects on the table.

He hooked his finger through the trigger guard of her pistol and lifted it up. "Now why didn't you use your gun to shoot my guard? Afraid the shot might warn us as to your arrival? And this"—he picked up her earpiece—"why the need for communications, if you're here alone?"

She gulped, thinking of several glib answers but not sure how wise it was to use them. He didn't wait and held up the watch. "And what will I find on here?"

Not much, as it had undoubtedly fried in the power surge. But she could play along. "The five nearest airports?" Mike would be proud of her answer.

The more Brugman talked, the less he scared her. He had paled slightly when he lifted her gun. She guessed he never did his own dirty work and avoided the necessity as much as possible. He pushed her ankle knife around on his desk but didn't pick it up.

209

"I don't tolerate people who lie to me." He glanced up to search her face for reaction.

"Is that what the guy did whose leg is left in the chain? I'd heard you didn't stoop to those low-life actions."

Brugman's face reddened. "You're the one who's amateurish." He deflected the conversation back to her. "The young boy in your care was simply handed over to me. I told Kagona, Taljaard would come to get him back. All I had to do was wait, and you walked right into my trap. Now I'll wait again for Taljaard to ride to the rescue. It's all too easy."

Muscles spasms shook her frame, and her fingers and toes tingled as they returned to life.

Mistaking her spasms for fear, Brugman smirked and rubbed his hands together. With a touch to the front of his desk, a painting of Victoria Falls slid aside, displaying two security monitors. Brugman scooped up a remote and punched on a screen.

He zoomed in on a dark space. A single light shone overhead. Joni's heart clutched. Sipho sat on what looked like red plastic milk crates. Cloth had been thrown over the contents of shelves behind him, but the edge of what looked like a cereal box was visible. Food storage was near the kitchen. Is that where Brugman kept Sipho?

"I'm able to utilize my men outside hunting for Taljaard while I monitor the boy from here. He's quite secure. It'll cost Taljaard time and energy finding the boy. Even if he manages to invade my lodge, I have several things to slow him down. While I don't like violence, I am a man willing to protect his property from thieves or even terrorists. You never can be too careful these days."

Brugman's self-aggrandizing rant went unnoticed. She focused on Sipho, who folded one strand over another on the braided toy Ian had given him. He stopped and stared straight ahead. Long streaks glistened from his eyes down his cheeks. He'd been crying.

"You'll have a long wait. Taljaard isn't here." Vehemence filled every word. She had to regain control.

"So you say. I shall soon find out. Every door on this complex is monitored. Each time one opens or closes, the computer automatically switches the monitor to that room to pick up the person entering or leaving." He touched another button and each screen split views. "I did mention I had armed visitors, didn't I? Even if Taljaard knows where the boy is being kept, he won't get far."

"Don't be so sure. I strolled into your place from the nature sanctuary. Do people know a crazy man runs this lodge?"

"Crazy, no. Dangerous, perhaps, but I'm a realist, too." Brugman zoomed the camera to Sipho's leg where she could see something shiny around his ankle.

"You'd better tell me that bracelet is only a tracking device."

"It isn't lethal as long as no one tries to take it off or move him out of the room. It requires a code to disarm it and the proximity trigger."

"You could kill the child." For a second, Sipho glanced toward the camera as though he knew she was watching. If so, he looked quite unhappy at her failure. *Hang in there, kid. Help is on its way.*

"I won't hurt the boy. He's worth money to me. Of course, once I turn him over to Kagona, I can't make promises." A red light blinked on the monitor console. Brugman bit his bottom lip. "And it appears he has finished dealing with the boy and is ready to meet you."

Joni didn't see anyone in the room with Sipho, but she recalled his quick, disheartened glance toward the camera. Someone had been standing under the place where it was mounted. Evidently Kagona preferred to remain out of sight.

Relief flooded through her as she recalled Sipho's glance. She guessed Kagona's interrogation had netted nothing, and that probably wouldn't sit too well with the intelligence chief.

She straightened in the chair. Chalk one up for the kid.

Brugman collected his monkey. The automatic door slid open. He stopped and scowled back at her. "Don't expect Kagona to be as nice."

The door whooshed shut behind him.

CHAPTER TWENTY

Five minutes later the guard at the door left. Joni stared at Sipho, still working away at his braid craft. The monitor flicked off, and the painting slid back into place. Brugman playing games?

The room lights dimmed to a mere glow and eventually left her sitting in darkness with only the red numbers on a digital clock lighting her space.

Cute psychological trick. Make her hope they had forgotten about her. Like Brugman would forget a wad of cash. A small spotlight high on the wall behind the desk popped on and illuminated a carved rhino perched on a pedestal. Light gleamed off the polished wood. Slowly the illuminating circle moved off the figure and appeared on the chair back, sliding down the front of it like a kid on a slide. An arc and then a full circle of light moved onto the desk. The plate-size spot inched across the desk, adjusting as it moved to line up with her position.

Joni pushed up enough on her feet to slide her chair out of its line. The circle altered its course. She scooted farther in the same direction. Once again the light adjusted. Hell, she'd move all the way to the door if she had to. On her next attempt, the chair caught on the edge of something and wouldn't move. She jostled harder, to no avail. Her pulse raced.

The spot, now an oval, reached the edge of the desk and flowed onto the floor a foot from her chair.

"Just shine the damn thing on me."

The beam hit her legs and flowed onto her lap. She turned her face and eyes away, hoping it would pass. The light stayed.

"I happen to know Ian Taljaard isn't married." A deep, muffled voice came from near the desk.

She hadn't heard the automatic door slide open, so whoever spoke must have entered from somewhere else. Keeping her mouth shut at this point seemed like a wise idea.

"Yet some men working in his town claim you are his wife. What is the truth? Are you or are you not Mrs. Taljaard?"

Only one man had a network that could garner information so quickly about their run-in with the roadblock gang. The voice belonged to Kagona. He operated the state intelligence organization and no doubt had the upper hand in this interrogation. Her best defense was to sound so convoluted, he'd give up on her. Yeah, when crocs flew.

"Are you finding it hard to answer?" he prodded. "I'll introduce myself if you tell me your name."

A one-sided conversation gave him little to work with and would probably lead to physical action to encourage a response. She'd had enough physical for one night, so she turned her face toward the beam.

"Take the light off my face and I'll think about your offer." She did a lousy job, faking an accent.

A few seconds later, the spot went off and the room lights came on. Two men occupied the room. One, who was NFL size and wore a loose shirt with enough room around his waist for a complete arsenal, was likely a bodyguard. The other, settled against the edge of the desk, signaled him to guard the main door.

"I have met your request."

The man before her must be the notorious Kagona. Tall and robust, his skin was a rich brown, but not a heavy dark tone many Africans bore. His face appeared a bit swollen and perhaps even a shade darker on one side by bruising hidden under a thin bandage that stabilized his jaw. The bandage wrapped from the top of his head under and around his chin and then was secured at the back of his neck.

Fierce eyes narrowed at her assessment of his injury. "What do you have to tell me?" He spoke through a clenched jaw. Kagona would be out to settle the score for Ian's handiwork. Hopefully he'd hold off until Ian showed.

She shrugged, finding new places that ached. "Several things come to mind, but my parents taught me not to say anything if I couldn't be polite."

"And where do your parents live?"

"I've been taught not to talk about family to strangers."

"We won't be strangers once I have your name."

Buy time. "Since you say I call myself Mrs. Taljaard, that will do fine. And you, sir?"

Kagona's stony mug masked his irritation, but his fingertips—dying to drum the desk—twitched. He straightened and positioned himself directly in front of her chair. His height and large frame made him quite imposing. He crossed his arms, revealing a long scar on the inside of his arm. This guy lived a violent life.

"Tinashe Kagona."

"My memory's a bit fried. If you don't mind, I'll call you Kagona."

The backhand slap came hard across her face and left her tasting the iron of blood in her mouth. Evidently, he minded. Back to keeping her mouth shut.

"You're a spy and should be shot as such. Outsiders have no right to influence the future of this country. Who do you work for? Americans? Aussies? Brits? Perhaps the Canadians?"

She hung her head down, not so much from pain, but to avoid answering questions. It didn't work. He grabbed her chin and squeezed as he lifted it up. Pain shot through her jaw. For a second, she wondered if he intended to break it like Ian had done to him.

"I'll ask one more time." His request huffed out quite well through clenched teeth.

"Canadian, but it doesn't matter." At least the story would waste time. "I'm a freelance reporter. My focus is on uncovering government corruption in Africa. Your opposition will love this story."

He released her jaw, as though satisfied. "Interesting." He collected a tissue from behind Brugman's desk and wiped at his hand. "Do you commonly camouflage your face and invade private property?"

"Uncovering corruption is a dangerous job. I do what it takes."

"I see." He walked behind her chair.

She willed her head not to turn. Fear threatened to shake loose, not being able to see what he might do to her.

"Perhaps I could trade you back to Canada…if that is your country." He paced back beside the desk. "Although it is much easier for you to die miserably, another hippo fatality on the Zambezi. Whatever your country, they will learn not to send more operatives into my territory."

The picture wall opened again, and the monitors reappeared. Brugman probably hid safely away from the violence in a central control room, where he operated the electronics. Sipho appeared on one screen, still huddled on the plastic crates. Ian hadn't made his move yet.

Since Kagona had stepped out of hitting range, she felt safe tossing out an insult. "People who lock children up are on a paved road to hell."

"Brave talk. You and Taljaard cannot get the boy out safely. It's impossible. I've followed your progress since you left Bulawayo in a helicopter. You landed at the Victoria Falls Hotel. Does this look familiar?"

Another monitor switched on. A photo appeared of two men standing in front of Mad Mike's Jet Ranger. The sun behind them had not yet set. One man held up a vital piece of the chopper.

"A new battery is hard to find. One would have to be flown in. That could take days, weeks, maybe never. Adventure Tours will have no future without a helo."

Losing the chopper came as a blow. Without it, the trip back to Taz after rescuing Sipho would be a long and arduous prospect. But Ian had said he planned for contingencies. A disabled chopper counted as one...she hoped.

"There are only two easy ways into this villa, by air or river. Since you didn't arrive by air, I assume you came by the river."

"I walked from the nature sanctuary."

"Another lie. That journey is treacherous. Brugman feeds wild dogs along his border with the sanctuary." Kagona shook his head and moved to stand before her. "Brugman will find your boat. His men search for it now. I foolishly told him he wouldn't catch Taljaard at the dock. Imagine my surprise when he pulled you in."

"Shocking," she said unable to help herself. When circumstances became so unbearably bleak, humor helped. So why didn't she feel more confident?

"You won't laugh when I decide what to do with you." Did a twinge of anger edge his voice? Now who was the one getting all worked up? "I can't imagine how you injured my pilot and his crew members. Whoever sent you, trained you well."

An explosion shook the villa. Excitement flickered in Kagona's eyes. "Ah, the fun begins. Are you ready to watch Mr. Taljaard fail?"

Smaller booms, rather like fireworks, echoed against the

walls, coming from more than one direction. Mad Mike must have joined the melee. She waited to hear the rattle of automatic gunfire, but none came.

"I want Taljaard alive," Kagona said as if he could read her thoughts. "There is more you can do to a man than shoot him. My men have orders to ignore the diversions and remain close to the lodge."

Her gaze flickered to the monitors. Sipho held his hands over his ears, looking distinctly grumpy. As she had witnessed, loud sounds bothered him. *It'll be okay, kid. Ian promised he wouldn't fail. I believe him.*

Did she? What were two men's odds against two maniacs and who knew how many armed thugs? Their single objective was to capture Ian. Ian, on the other hand, had tunnel vision, with his one purpose to rescue Sipho. Even if he figured out all of Brugman's petty traps, to defeat them required time. It didn't take an expert to see he didn't have much left.

Her capture, advertising their presence, hadn't helped. Ian must have hoped to divide up the thugs by attacking several locations, but it didn't appear to be working.

Kagona ordered the outside cameras on. Nothing happened. "Go see what's the matter." The bodyguard left the room, and Kagona drew the pistol he wore at his waist.

A minute later the man returned. "Brugman say to work monitors. He go look at damage by Taljaard."

Kagona snatched up the remote and switched the scene to outside. Save for light emanating from the windows in the lodge structures, the lawn and courtyard were dark. Ian had taken out the lights. Joni held back a smile as Kagona stabbed at the controller keys, increasing his force and frequency as whatever he wanted didn't appear on the screen.

Evidently Brugman's men had not secured the grounds as well as Kagona preferred. He threw the remote at his bodyguard. "Get me the infrared, now."

The man played with the remote for a few minutes until the monitors turned hazy. They stayed that way until a glowing human form appeared on the screen. Several more joined that one, no one necessarily following the others, but rather moving haphazardly. It was impossible to tell who belonged to Kagona or Brugman and who might be Ian or Mike.

Another explosion sounded. The monitors and lights flashed off. How many men did Brugman or Kagona have guarding the vital power source? Two? More? Joni's spirit buoyed.

A close-by generator started up. Lights and electronics came back on, but the monitors stayed blank. Kagona snatched back the remote. A monitor flickered on.

"You got your wish for action." She might as well ratchet up the pressure already flushing his neck red.

No matter how hard he punched at the buttons, only the video of Sipho came up. The exterior shots remained blank.

"It's tough when you can't see what's going on out there." Joni struggled against her restraints. "There's always the window." Which was behind heavy curtains.

Kagona snickered through his teeth at her efforts. He tossed the remote on the desk and leaned back against it, doing his best to look unruffled.

That's right, bastard. Focus on me. The boy still had his ears covered, but on that chin he had tucked to his chest, she swore he wore a smile.

Optimism surged. "I hope you saved your best man to keep you safe."

Kagona smiled at his bodyguard. "Hope all you want. I know where to put my best defense. Time has run out. Taljaard must act soon."

He picked up her ankle knife Brugman had left and slid it from its sheath. Holding it up to the light, he gently tilted it back and forth before running a finger near the edge. "A dangerous thing for a woman to carry."

Excitement or tension, she couldn't tell which, tightened the big man's shoulders. With a flick of his wrist, he sliced open the leather on the desk chair. Brugman wouldn't be happy.

"Sir." The bodyguard pointed to a communications piece in his ear. "The men have caught Taljaard near the kitchen."

The kitchen abutted the food storage. Right where Ian should be. Despair coursed through her as she glanced back at Sipho. She had hoped his cage would be empty.

Kagona caught her glance at the boy. "Taljaard's informant had passed along the boy's location at food storage as predicted. We were ready for him." With great force, Kagona stabbed the blade into the desktop, making her jump.

Once again unable to prevent death or injury to an innocent child, she dropped her head, resigned to the fact that all Ian's promises had come to naught. Or had they? Some strength inside brought her chin up. Commanded her to look at Sipho.

Her faith hadn't died. This mission wasn't over. Mad Mike was still out there somewhere. From how these men boasted about working together, they didn't believe in surrender. Mike wouldn't give up any more than Ian, and that left room for hope.

A commotion outside the office drew Kagona's attention toward the door. She craned her neck to see as three men struggled to bring their hostage into the office. Kagona's face tightened as the men entered.

They pulled the hostage upright. With blood dribbling from his mouth and purple bruises already forming on his face, he stared straight at Joni.

"Good God" slipped out of her mouth.

"Fools." Kagona slammed his fist on the desk. "That's not Taljaard." He gave orders in a tribal language, gesturing wildly to make up for his inability to shout.

Mad Mike grinned at Joni as the minions bound him in the slashed desk chair. Brugman's men aimed a gun at her to assure his cooperation. Once done, Kagona sent away the men to find

Taljaard, retaining his bodyguard and one of Brugman's men.

Kagona tempered his rage. At least she had a moment's reprieve while he focused on Mike. "So you're the Adventure Tours pilot. It was a foolish thing to help Taljaard."

"Taljaard?" Mad Mike had the nerve to shrug. "No, sir. I work alone, helping ladies in distress."

Kagona, his rage reheating, snatched the knife protruding from the desktop and stepped toward Mike. "I am tired of tall tales. Where is Taljaard?"

"In Harare," Joni answered. Ian had said to buy time if they were ever captured. Before one of them became a kebab on Kagona's skewer, now was the time for purchasing.

Both men focused on her with questioning eyes. She did her best to look uncomfortable. That was relatively easy, considering her hands were tied behind a tall chair with a maniac about to slice and dice her latest acquaintance.

"He knew Brugman also has a home there," she added. "He figured you'd prefer to meet in your own territory and make Brugman bring the boy there."

Kagona pointed the knife at her. "He was seen with you in Bulawayo."

"He secured me a helicopter and this man to help, because there was a chance Sipho might be here. We had to split up and cover both places."

Kagona looked to his bodyguard. Without saying a word, the man slipped out the door. Was he checking with his men who were at the Victoria Falls Hotel to find out if Taljaard had arrived on the Adventure Tours helicopter?

He redirected his attention back to her. "You're lying. Whoever you work for wants the boy. Who is that?"

"No one. I came to see Ian. I knew him from my university days."

"A good number of years have passed since he attended school. Why now?"

"I figured he'd be a good contact to help me sniff out corruption. Considering I'm sitting here, I'd say it has worked well."

"What about the boy? What is he to you?"

"He's a friend of Ian's. Nice kid. I told Ian I'd help find him. That's why I'm here."

Kagona's bodyguard returned and held up three fingers.

"You know no truth." Kagona looked irritated, and likely in pain. All this talking couldn't be good for his jaw. He signaled his bodyguard over, and they spoke in hushed tones.

Once his guard returned to his position, Kagona slowly paced before his guests. "Did you think that little trick would buy you time? Three people landed at Vic Falls. You rode up front with this pilot instead of Taljaard. Why?" He stopped before her. "Are you a pilot, too?"

Mike laughed. "Wrong. A tourist paid me a few dollars for a ride. Wanted to sleep in back. I told this lady here, the best view is up front. We'd be over the falls if she had touched the controls."

Kagona ignored him and hovered behind Joni. "Reports came to the police of a strange craft flying near Bulawayo. Could you have flown a plane to Zimbabwe from South Africa? Tell me where you hid the aircraft."

"The Jet Ranger is at Vic Falls. Unless you mean the helicopter I saw sitting outside. Yours?"

He signaled his bodyguard. "I want these people motivated to provide answers. I have the boy. If Taljaard wants him, let him show his face."

Kagona's bodyguard called in two extra men.

"I told you he's in Harare."

Kagona leaned inches from her face. "Taljaard is here. He was seen at Vic Falls." His breath, rank and heated, blasted her face. "You will tell me his plan and where your aircraft is located, or you will die."

"I can't tell you what I don't know." That was true. She'd no idea what Ian planned.

"I know what you fear. Brugman caught you because you were afraid to be near the water. Did the leg in the iron at the dock frighten you?"

While the horrid vision of human leftovers filled her head, she did her best to remain calm. By the pleased twinge at the edge of Kagona's mouth, she failed.

"I want answers from them. Find out if Brugman's pets are still hungry."

Fear welled inside. She'd imagine many ways to die, but croc fodder wasn't one of them.

Kagona stalked to the door but stopped for one last parting shot. "Whoever talks first, lives."

He pointed at Mike. "One wrong move on the way to the docks and her throat is sliced."

The bodyguard followed on Kagona's heels, leaving three men in the room, two with big guns and one with a long, throat-slitting knife. The best she and Mike could hope for was a rescue on their way to the feeding frenzy. Being the main course did nothing to offer confidence in surviving until Ian's main objective—rescuing Sipho—was met.

A man lifted Joni straight up out of the chair so he didn't have to retie her hands and shoved her across the room. Amazingly, her legs still worked. She assumed her arms did, too, but the rope had long ago stopped any circulation. As she stumbled past Mike, he gave her an encouraging wink. What was that for? Did help wait around the next corner?

The entourage shuffled along the dark path on the way to the dock, and at every twist and turn she expected Ian to show. So did their escorts, considering the men's itchy fingers on their weapons. They were almost to the river and so far nothing. Nada. Where was Ian, her knight in shining armor? Surely Mike's wink meant he had a plan or Ian had a plan or someone had a freaking plan, because she surely didn't. Being priority number two sucked.

They arrived at the dock much too quickly. Darkness bathed their outdoor torture chamber, as the light that hung on the wood pole was off. She had a vague recollection of seeing its last flicker of life.

Perhaps the men would forgo the ankle chain and simply dip them into the water to tease the crocs. One guy picked up the shackle lying on the sand and shook the leg free. Then again, perhaps not.

"You work for Brugman, right?" She spoke in desperation to the man shoving her toward the shore.

He didn't answer.

"I can get you money. People will pay for us. It makes good business sense to keep us alive." Her groveling made no dent in his armor.

The leader of their merry band shone a flashlight into her face and then Mike's. So much for the little night vision she had gained on their stroll.

This guy knew his business. His stern mouth had remained clamped shut the entire trek. Brugman paid him too well.

He pointed toward Mike, and the men hauled him over. Mike cussed and struggled against them as they tried to clamp his leg.

"Neither of us are going to talk," he yelled. "Why don't you just shoot us?"

Joni frowned, reconciled to an unpleasant fate but chagrined at Mike's request. Why invite a quicker end than dinner by starlight?

"Sorry, missy," he added.

She considered his actions. Ah, heck. Mike didn't give a damn about being eaten alive or shot. He was up to something.

"Crocs like fresh meat," one man said as another succeeded in clamping the leg iron to Mike.

There was only one chain and iron. Did that mean she got to watch? As if on cue, the man with the light dug around a footlocker on the dock and brought out another leg bracelet.

Yep, she and Mike were auto-rotating to a hard landing.

They hauled her next to Mike, clamped on the leg iron, and left her sitting in the sand. With large pliers they clamped the other end of her chain to an iron loop on the metal pole.

"Don't worry, missy. It won't be too bad." Mike sounded so *not* reassuring.

"Yeah, the croc just had dinner. Maybe he won't want a midnight snack."

"Fat chance of that. Crocs are willing to share territory in places where they know food is plentiful. He'll have a few friends waiting for their meal."

"You always such a pessimist?"

Mike grinned. "Only when I'm about to die."

She worked close to his head and whispered, "Do you have anything to pick these ankle locks? I can work my hands free."

Metal clanged as Brugman's man tossed the pliers back into the locker. He hopped off the deck that stood elevated about waist high above the shore and shuffled through the sand to stand before them. "I have a key. First one who tell me what Kagona want, I let free."

"And the other?" She then thought better of hearing the obvious. "Never mind, don't answer that."

Somberly she watched as one man collected the bloody limb. The rest extinguished their flashlights.

She remained silent, listening for footsteps to tell where they had gone.

"How are you going to hear if we confess?" she yelled.

A man replied from the dock, barely fifteen feet away. "I hear perfect. You ready to talk?"

"Fuck off," Mike retorted.

A splash, likely the remains of the former sacrifice, came

from the end of the dock. One man started slapping a paddle on the water while others tossed rocks off the dock.

"Is that what I think it is?" she whispered to Mad Mike.

"Yep—the dinner bell."

CHAPTER TWENTY-ONE

Kagona leaned against a chair in the main lodge and spoke via radio with his man at the dock. No capitulation yet from the prisoners.

The resort doors opened and Chaipa barreled in. "I spotted Taljaard at employee quarters. But Brugman's men found no one there."

"You're positive it is Taljaard this time?"

"Yes, sir."

"Everything the woman claimed about Taljaard was a lie." Kagona should have taken a quick photo of her to send to his contact in South Africa and confirm whether or not she was the mysterious aircraft pilot. A description would have to do. At least natural redheads were uncommon.

"The woman is an amateur." He typed a message into his phone. "She will break before being eaten alive."

The woman had failed. His enemies had failed.

The Shona were the rightful rulers of this country. They had the right to its riches. His people had built Great Zimbabwe a thousand years ago. Zimbabwe would again be the greatest power in Africa.

A rustle behind him brought his pistol from his waist holster as he spun around. Brugman's damn primate scooted across the room. Tempted to shoot, Kagona stuffed the gun back in the holster.

Taljaard was close by. He could smell it, feel it in his pulsing jaw. Revenge swirled in his blood. How many of Taljaard's bones could he break and still keep him alive?

Kagona recoiled, remembering his humiliation when an old widow discovered them in the farm well. She had taken an eternity to bring relatives back to get them out.

Taljaard would not die quickly for that disgrace. Kagona's men had orders to bring him in alive, and Brugman would get no money for Taljaard if he died. Taljaard had enough smarts not to walk into any of Brugman's stupid traps. Kagona smiled. He had set his own. He wanted Taljaard humiliated in failure, tormented, and then dead.

Kagona turned back to see Chaipa and his bodyguard staring oddly at him. "Show me exactly where Taljaard's Adventure Tour pilot was captured." Chaipa had been the one who caught him.

Chaipa led him down the breezeway connecting the main building to the kitchen and storage facility. He stopped before entering and pointed to the thatched roof, which had a large hole in it. "The pilot dropped smoke canisters into kitchen and storage room to block security cameras. We catch him in kitchen on way to storage room."

Kagona studied the hole and walked around to the door leading to the kitchen. "How long between the smoke bomb and his capture?"

"Fifteen seconds, sir. Maybe few more."

"Enough to jump from the roof and head inside." Barely, but it could be done. Did the pilot plan to enter the storage area alone or was Taljaard with him? "What did you find on the man?"

Chaipa shrugged. "A knife and a lock pick."

"No gun? No explosives? Comm gear?"

"No, sir. But he disabled two men before caught."

"That old man?"

"Smoke was in kitchen. Then power went out."

"Sounds as though the pilot was expecting those events. He was a diversion." Kagona nodded at his bodyguard. "Taljaard never went into the storeroom. He knew the boy wasn't there."

Kagona ran back toward the main lodge. He drew his weapon and bypassed Brugman's office for the electronics room, which housed security. His bodyguard and Chaipa, carrying an AK-47 at the ready, stayed with him.

Kagona tried the reinforced door and found it locked. "Damn Brugman. He trusts no one. Through the office." Kagona had entered from the monitoring room via a secret panel to interrogate the woman.

The panel opened to pure chaos.

Brugman's technician lay bound on the floor. Much had been damaged, but oddly the video feed showing the boy remained intact. He sat with his hands over his ears and hadn't shifted positions since Kagona had last seen him on the monitor in Brugman's office.

"I want Taljaard dead." Kagona's vehement pronouncement stretched his jaw against the bandage and shot pain in a dozen directions into his head.

Hate filled him at the simplicity of his enemy's moves. The video feed had been frozen. Still, the boy couldn't move past the door with the anklet activated.

Kagona signaled Chaipa to open the equipment closet. Inside, he'd staged it as food storage in case Taljaard hacked the security feed.

Chaipa slammed open the door. A single light overhead swung back and forth. He shook his head.

Kagona stared into the empty space, unable to digest the truth. On the floor next to the milk crates lay the anklet Brugman had claimed would keep the boy in place. Kagona swept up a plastic crate and swung it at a monitor showing the boy. The screen crackled and the picture blinked out.

An explosion sounded in the distance. Kagona glared at Chaipa, who had a finger to his earbud.

"Sir, Taljaard blew up sleeping quarters. Brugman ordered his men to protect central lodge. He's afraid Taljaard will destroy this place next."

"Brugman's a fool. Taljaard already has."

"Phew, plue." Joni spit sandy dirt from her mouth as she knelt behind Mike in the dark. "You could have kept the damn ropes clean," she whispered, knowing a man still stood up on the dock after the rest had run back toward the main villa. "Think that guy was left with NVGs?"

"Doubt it. I only noticed one with them. Ian is the biggest threat. Those hunting him will hang onto them. Besides, if our guard was watching, he'd be telling us to separate."

A light shone on them. "Ready to talk?" the guard asked.

"Hell, no." She spit out more sand. "Not to the likes of you."

The guard laughed and clicked off the light.

"Quit talking, missy, and keep trying," Mike whispered. "Brugman could damn well change his mind and order his men back any minute." Water splashed nearby. "Hurry, the wildlife is stirring."

A shadowy croc emerged onto the sandy shore. Meager moonlight glinted off its jagged, protruding teeth. It dropped to its belly at the water's edge. Reflective eyes watched them with interest, trying to determine who would make the tastier meal.

Frantic, she clamped down on the nylon rope and pulled. The smell of damp earth, acrid wet reptile, sweat, and pure fear permeated the air. After pulling with her teeth so many times that she'd lost count, Mike's rope loosened.

"Just a little more." She spit again before renewing her efforts.

After several productive jerks, the end pulled out. "Good to go," she whispered, hardly able to control her excitement at the small victory.

Just in case the guy put them in the spotlight again, Mike let the rope hang loose at his wrists and went back to back with her.

"It's going to take forever this way," she groused.

"We can't be certain whether the guard can see us."

What seemed like an eternity took merely seconds.

"You're good." She slipped the rope into her hands but kept her wrists behind her back. Another pair of long slits glistened in the moonlight near the water. "Looks like Dad invited the whole family to dinner. What now?" She had backed as far away from the water and encroaching crocs as possible.

"Find out where the guy is on the dock."

"Hey, Brugman's buddy," she yelled. "The crocs are coming. Are you close enough to get me free if I start talking?"

"I am here," the man answered from the deck of the boat.

She worked her way back to Mike. He stood behind her and yanked down his pants. "Cover me." He plopped on the ground and pulled his pants past his knees.

Puzzled by his actions and unable to see clearly, she scooted closer, blocking him with her body. "What the hell are you doing?"

"Saving our lives." Mike played with something wrapped on his leg. He lay back in the dirt and hiked his pants back on. Before she could ask how lying down on the job would help, he rolled over to his stomach.

With surprising agility for a sixty-year-old, he pushed to his feet.

Sheesh. She looked again as he hopped unsteadily. He had pushed to his *foot.* Left in the iron ankle bracelet lying in the sand was his prosthetic foot.

She stepped up so he could use her for balance. "That's what all the struggle and cursing was for when they chained you."

"Had to make sure they clamped the correct leg."

"You going to be okay?"

"Nothing I can't live without."

"Fine for you." She kept her concentration fixed on the two reptiles positioned along the shore. "But I won't be alive much longer if you don't get me free."

"Help me toward the side of the dock. Then draw the guard over." Mike hopped toward the raised dock, using her as a crutch until the chain brought her up short. He made it the rest of the way in the loose sand and dirt and bent down against one of the large floating drums.

Slowly, so as not to further excite the crocs, she backed away.

"Okay, Mr. Guard," she yelled again. "There're two crocs. They're getting closer. I'm ready to talk. Now." She threw in enough panic to encourage the man and not the crocs.

Boots clomped closer on the deck. Mike needed him at the very edge.

"Hurry, for God's sake. I don't want to be torn apart." The crocs sensed fear in her voice and moved forward. She toned down her panic. She'd seen the leftovers from their last feeding and didn't care to be the next course.

The guard lurked on the dock, still away from the edge.

"Hurry, or it will be too late for me to say anything. Kagona won't be pleased and you won't get paid."

"Give me something to tell Kagona. He decides." He still stood several feet too far down the dock for Mike. She could see the guard's outline, which meant he could see hers and not Mike's at the pole. "Where is the man?"

"He buried himself, thinking the crocs wouldn't dig him up." She had to move closer to the crocs to draw the guard to Mike's position. Her body shook. Damn fear. "Show me the key before I say anything. I want to know those other men didn't take it with them."

"You talk first."

"Son of a bitch. You'll have nothing if the crocs grab me. It can't hurt to show me the key."

She could see the man hold something up. "I can't see it. How do I know you aren't holding a paper clip?"

He stepped to the edge of the dock. Mike had him facedown and unconscious in the sand before she could think of anything further to say. He grabbed for the guy's hand.

"Are you hurt? Bring me the key."

"Can't find it."

"Shit. You lost the damn key?"

"Don't blame me, missy. You're the one who told him to take it out." Mike lifted himself onto the dock as the two crocs spread apart on the shore. A third one emerged to take the middle position. She hated teamwork.

"Mike, get me his gun."

"No way. Using it would bring everyone down on us. Don't worry."

"Easy for you to say. You damn well better be hunting for something to set me free. Hurry, for God's sake. The gang's all here and ready to eat."

She heard Mike roll the few feet to the footlocker, where Brugman's men had stored the pliers. He scrounged through tools in the dark. A second later, he rolled back and dropped off the dock onto his good leg. He used something as a crutch to move closer until he reached her and headed straight to the pole.

The crocs correctly interpreted his movement as a cancellation of their dinner reservation and started to approach. Mike swung a crowbar at the metal loop on the pole securing her chain, breaking it loose. The noise brought a momentary halt to the croc advance. He locked an arm around Joni's shoulder, and they barreled toward the dock.

The reptiles raced after them. Joni leaped up and landed on

the dock edge, the trailing loose chain still fastened to her ankle. She swung around and caught Mike's shirt in her hand as he pushed off the sand, pulling him up with her as the crocs sprang forward. He rolled with her to the far side of the dock with crowbar still in hand. Something caught her ankle and stopped them short of rolling off the other side next to Brugman's boat.

They both looked back, fully expecting the crocs to follow. They were alone, she underneath and Mike on top. Both their hearts pounding into one another.

She laughed quietly, but hard. "Sorry, Mike, but I usually insist on dinner first."

"Blimey, I save your bloody arse from being a main course and that's my thanks."

She took his head in her hands and planted a big one on his cheek. "Thanks, now let's get outta here before the rest of Brugman's men return to pick over our remains."

A horrid scream cut the air, followed by frantic cries in a native language. Mike rolled off her with the crowbar ready. They peered up to see where the crocs had gone. Two sets of gleaming eyes stared back. The other had clamped onto the no longer unconscious guard.

Joni yanked on her ankle chain, realizing it was what had stopped her and Mike from joining the extended family of crocs on the boat side of the dock. The chain held firm. She yanked again.

"Eh, missy, I wouldn't do that." The concern in Mike's voice filled her stomach with dread as the screams of the dying man accosted her ears.

As she looked into the shadows on the sand, a set of eyes stared back, and the reason for Mike's tone became clear. Moonlight glinted off the metal chain caught in a croc's mouth. He had clamped down and was backing away with her in tow.

It didn't take a genius to see who would win this wrestling

match. After her last tug, tension tightened the chain. It pulled her ankle back toward the shore edge of the dock and her waiting friends.

The guard's screams ripped the air. The planking, slick with sand and water, helped her to slide toward the edge.

"Mike, I need help here." Desperation sent her clawing for any handhold. As she slid by Mike, she yanked the crowbar from his hand. Lots of good it would do against the bone and leather of a croc's head, but it was better than nothing.

Mike scrambled toward the footlocker.

Joni shoved her other foot against a metal cleat on the decking near the same pole that had earlier broken her fall and battered her body. With all her might, she pushed to hold firm.

The croc paused, perhaps unsure as to what had stopped its easy progress. She used the moment to swing around to the opposite side of the pole, letting the chain wrap around it. Teetering on the edge of the dock, she wove the crowbar over and under the chain length on one side of the pole, with the other side heading toward the croc. She twisted the crowbar once as the croc yanked.

The chain went taut, but the crowbar locked it against the pole. The links strained and creaked, ready to break loose. Pleas of frantic help carried across the sand. She didn't want to add her voice to the chorus.

"Mike!" She grabbed onto the pole.

"Keep your head down." Objects whirred past and landed among the crocs.

The chain loosened, so she yanked. The crowbar dropped to the ground. Freedom. She reeled in the rest of the chain before crawling over to the footlocker and helping Mike empty its contents. The screaming ceased, and they heard the splash of water as the reptile family left.

The buffet had turned into takeout dining.

She gulped, attempting to collect her shattered nerves. Her

heart pounded faster than Zulu drums. "Where to now? The rendezvous point?"

"To the boat first, then I'll send you on to rendezvous with Ian."

"Hey, I'm prior military. I leave no man behind."

"You're not leaving me behind, missy. I've some business to take care of with the boat." Mike rolled to the dock edge and dropped to his foot.

Joni followed, looking for the crowbar and keeping her eyes peeled for any return visitors. After a few shuffling passes through the soft sand and dirt, the bar struck against her boot. With a few strikes against the metal pole, the chain holding Mike's prosthesis was free.

She dumped dirt from the prosthesis cuff and handed him back the leg with attached iron hardware. "How are we going to get these things off?" She stuck out her ironclad ankle and twirled the end of the chain in her hand. "Swinging a crowbar at my ankle has dumb ass written all over it."

Mike divested her of the crowbar and tossed it in the sand. "You know, missy, you're starting to pick up a sense of humor." He wrapped an arm around her shoulder for support. "We have to move out. On the way back to the boat, we'll pick up my pack I hid away. I've something in there that should work on that bracelet."

He tucked his foot under the other arm and leaned heavily on her as he hopped over the softer ground. "Sorry, the travel won't be easy."

"You don't hear me complaining. It beats the hell out of digesting in croc stomach juices."

CHAPTER TWENTY-TWO

Fury simmered Kagona's blood. Brugman's technician could tell them nothing about where Taljaard and the boy had gone. Kagona shoved him away.

"Has the man at the dock answered?" he snarled at Chaipa.

"No, sir."

"Radio Brugman. I want backup. Then follow me. Brugman was a fool ordering his men from the dock."

Kagona barreled out the door. His bodyguard overtook the lead, and Chaipa brought up the rear. Each jolting step shot pain from Kagona's jaw through his head. He'd been a fool trusting Brugman's men could provide enough protection from Taljaard. He should have taken Sipho immediately and bested Taljaard another day.

Should have, could have ran through Kagona's mind as he hustled toward the dock. He wanted his prizes back. He didn't fear an attack upon himself—Taljaard had the boy and wouldn't risk confrontation. He might, however, attempt to free his friends. Taljaard's greatest weakness would never change.

"If Brugman expects any money, he'd better hope the woman is alive." Radios crackled with messages. Kagona looked back to concern on Chaipa's face.

Chaipa shook his head.

Footsteps scurried far behind them as Brugman's reinforcements rushed to catch up. Kagona signaled to Chaipa. "Go to where our men found Taljaard's boat. If he hasn't taken the boy to it yet, wait for them. I'll send more your direction."

His operative bolted off past the path to the dock. Even if Taljaard had launched onto the Zambezi, Kagona still had hope. He'd covered all contingencies. If he couldn't have the data in the boy's head, then it was better no one else got their hands on it, either.

Joni and Mike made good time to the boat, including the stop to pick up the pack. Exhausted from the near-death ordeal and being a crutch to a man with a good sixty pounds on her, she collapsed on the ground next to the boat.

Worried about voices carrying in the night air, she leaned close to Mike. "I'll help you get this thing to the water."

"It won't be easy, but this time she's going downhill." Mike fished out his night-vision goggles from the pack and readjusted them before he examined the boat. "Something doesn't seem right here, missy. You don't suppose Brugman found it, do you?"

She sat up, leaning on her arms behind her. "Kagona disabled your Jet Ranger. He or Brugman probably found this, too. You want to check it out while I rendezvous with Ian?"

"Yep, need to plug the bottom, too."

"Can I wash that out for you in the river?" Joni pointed at his prosthesis.

"Haven't you had enough run-ins with crocs for one day?"

"But you can't stay here missing a foot."

"Says who? I don't need a babysitter."

She understood why men got off on the Ranger-SEALs-Special Forces trip. Achieving a goal with people who could relate to danger and death built strange ties—more than friends, different from family, and better than any ole job.

"You find Ian," Mike directed. "Since he's not here, my bet is he is holed up with Sipho. Evading the enemy and hauling a child with you isn't easy. He's waiting for someone to guard Sipho while he watches the bad guys."

"Are you sure he'll be at the rendezvous?"

"Yep."

"Exactly where is this place and how am I supposed to find it?"

Instead of answering her question, Mike rummaged in the pack. He pulled out a cloth roll with small, inset tools she couldn't make it out in this light. "Put your best foot forward, missy." She did, and he picked at the lock on her iron anklet until it popped open.

"There you go, good as new."

"Thanks." She rubbed where the iron had bruised her skin. "I'm going to add a pick to my wardrobe."

"Better get a set. I'm down one to Kagona tonight."

In the meantime, Mike took the NVGs off his head and handed them over. "Take care of these. They're top-of-the-line. Makes Ian's pair seem like shit."

She put them on while Mike continued with directions. "Before Brugman, a die-hard Brit lived around here. He obsessed about his land. Marked his territory by building a rock wall around the property. Most of it was so far out Brugman has let it disintegrate. You'll pick up the remnants about twenty paces through there. Stay to the outside of it."

Mike pointed off into the night and proceeded to give her precise directions to the rendezvous, along with another gun. Considering the pack had shrunk substantially in size since the start of the evening, she figured it was his last weapon.

"This yours?"

"Had it for a good twenty years. I expect to get it back."

"What about you?"

"I've other options."

Whatever those options were, she knew better than to ask. "How long should I wait for Ian?"

"Five minutes. If he's not there, he's not coming. And, hey, don't bring them back here if it's not safe. I know other ways outta this place."

Plans changed, and to survive you changed with them. They'd all get out of this somehow. "I'm not leaving you behind. Kagona's men will be headed this way. If you can get that piece of junk into the water and make it out to one of the islands, we'll find you."

"Counting on it. That's how friends work. Now get going."

Heading out, Joni picked up the wall and started toward the rendezvous. With every sense attuned to the river, jungle, and Brugman's estate, she made good progress. In the back of her mind, she recalled a comment about wild dogs. Recollection said they were small in size but hunted in packs. Great.

Mike had seemed confident Ian and Sipho would be waiting. The question was why—pure faith? She could use a little of that at the moment.

The wall, made of loosely stacked rocks, veered sharply right. Off in the distance she heard movement, as though a herd of humans bounded toward the river.

Were Brugman's men racing to the dock? Had they captured Ian and wanted to watch the finale as the crocs ripped and chomped body parts? Or had Ian successfully taken Sipho and they rushed to secure the last two hostages?

Energy infused her step at the thought. Maybe Mike had been right about Ian making the rendezvous near the north side of the lodge. A smart place, as the action had shifted south...toward Mike.

She made a quick detour to get around an overgrown area of bushes that didn't appear natural. Mike was right. Brugman hadn't bothered to maintain the grounds in the outlying regions of his lodge. Creatures rustled in a growth beneath a stand of palms.

Along the back wall she encountered a cracked stone bench. She had reached her destination. She climbed across the wall into a stone garden. Vines, moss, grasses, and wildflowers had long replaced the original growth. In the strange green hue from her goggles, the entire place seemed a surreal ode to the past. Stone pavers created a wide circular path, and inside the circle a mound of vines covered something in the middle. According to Mike, it was a bust of Queen Victoria on a pedestal.

Joni crouched and listened, surveying the area slowly and looking for a way through the vines into the center. A series of stone benches around the inner circle provided the perfect place to hide.

Step by loose stone step, she traveled the circular path. She spotted vines that had been sliced apart and subtly repositioned. Fresh sap dropped from the cuts.

Someone tapped her shoulder. She swung into them elbow first. Her elbow was deflected and leg swept out. She landed on her back with a body crashing onto her chest. Her breath whooshed out.

"I see you've no ill effects from Brugman or Kagona," Ian whispered. "Do you always strike first and ask questions later?" He pushed back her NVGs.

She gulped in air. "Sure, when someone doesn't bother to say my name before touching my body." Her limbs shook fiercely from the adrenaline he'd pumped into her system. "Damn it, Tal—"

He pressed his mouth over hers and let out a *shh*.

She pushed his face away and whispered, "You scared the crap out of me."

"Had to. Brugman has infrared cameras that swing through every five minutes. They're about due."

"Why didn't Mike tell me about them?"

"Did you follow the wall?"

"Just like Mike said."

"Then you were probably safe."

She partially shifted from under Ian in order to breathe. "What about now?"

"The stone benches block our body heat from the sensors. We're okay if we don't stay long. I took out many of his toys, but didn't get them all."

"Do you have Sipho?"

"He's under a bench waiting for you."

Elation racked her body. "But Brugman. How—"

"You get us away from here and I'll reveal my secrets. Are you done with the questions?" He didn't wait for an answer. "Good." His mouth covered hers, and by God she responded.

He cut it off as quickly as it started.

"What was that for?" she said.

He laughed softly and lifted his face from hers. "Making sure Brugman's shock treatment didn't have any lasting effect. You've tricky duties ahead. Come on." He pushed away, catching her hand and taking her up with him to a crouching stand.

Flustered and fighting to get a grip on the changing environment in this assignment from hell, she leaned close to him. "You damn well better finish that sometime."

He squeezed her hand. "My pleasure."

Ian let out a series of chirps like she had been hearing on the walk here. An answering call arrived. His communication with Sipho. She felt like Jane around freaking Tarzan.

Ian pulled her back down to a squat, letting the mass of vines and stone shield them. "Any idea where Brugman's men are?"

"By the stampede I heard on the way here, they were headed toward the dock. They'll find the man they left behind gone."

"Mike's handiwork?"

"He got him started, but the crocs did the rest. Mike isn't that far away from the dock at our boat, and he's a bit incapacitated."

Ian grinned as if remembering some past event. "One limb short, I suppose."

She offered Mike's signature answer. "Yep." Her voice became serious. "I won't leave him behind."

"Sounds like he saved your arse."

She smiled at the memory of rolling across the dock with Mike to avoid the crocs. "More than once. That's how he lost the foot."

"Don't feel too bad for Mike. It isn't the first time in the line of duty, and if I know Mike, it won't be the last."

A rustle of vines produced a dark figure crawling on hand and knees. Joni smiled as Sipho wrapped his arms about her neck and gave her a warm hug.

She embraced him and rubbed his back. "Missed you."

He squeezed tighter and nuzzled his head into her chest but said nothing. Tiny fingertips dug into her shoulders. What had he endured from Kagona? The images of her travails would haunt her nightmares forever.

Ian disentangled Sipho from Joni. "Ready to get home?"

Sipho nodded.

Ian crouched and grasped Sipho gently by the arms. "Remember, whatever happens, do exactly as I tell you and keep quiet."

Sipho drew a zipper across his lips.

Ian handed over a nearly empty pack to her. She started to ask about the other one he had taken but refrained. The questions would come later.

"Do we make our way back along the wall to the boat?" She slipped the NVGs over her eyes.

"Our boat's been compromised."

"But Mike—"

"He'll figure it out. It's up to you now to save us."

Damn, she hated obvious solutions, but not as much as she hated going back into the mouth of danger. The ragged teeth of the crocs reminded her of Kagona and his locked jaw. She sure as hell didn't want to see him again, either.

"It'll be guarded," she said.

Ian gave her hand a squeeze. "Yes, it will."

CHAPTER TWENTY-THREE

Kagona breathed in relief seeing Brugman's boat still attached to the dock. Its white fiberglass hull sparkled in a sliver of moonlight bouncing off the river.

"Get the spotlight on the boat," his bodyguard yelled at the men arriving behind them. "Shine it on the shore."

Two men complied and hopped on the boat. As the bright searchlight flicked on and then swung to illuminate the dock area, Kagona swore. Tools lay scattered, some impaled in the sandy dirt and others simply flung onto the surface. Floats and boat bumpers, blood and bits of clothing lay strewn about.

The radio on his belt vibrated. With the volume low, he held it to his ear. "Speak."

"Their boat is gone. The pilot put into the river."

"Are you sure it was just the pilot?"

"No doubt."

"And the boy or Taljaard?"

"I did not see them." Chaipa sounded breathless.

"Was the boat rigged?"

"Yes, sir. Once in the water, he has ten minutes."

"Let him go and come back. We must find Taljaard and the boy."

Kagona hooked his radio back on his belt. Brugman's second in command approached, but Kagona waved him away. He had to think. Taljaard required transportation. His boat and pilot had left without him. With the Adventure Tours helo disabled back at Victoria Falls, Taljaard's pilot couldn't fly back for a pickup. The local companies with helicopters had been warned not to permit any flights tonight. So what was Taljaard's plan?

He stared at Brugman's boat. There were only two vehicles on the grounds at this very moment—the luxury boat and the Alouette. Brugman claimed he could disable his boat from afar. Kagona wouldn't take the chance Taljaard could figure a way to disarm that feature.

"You." Kagona pointed at two of Brugman's men who carried AK-47s. "Stay here with the boat."

Taljaard alone could hike out through the nature reserve. But with a woman and boy in tow? Difficult and slow at best. Kagona's tracker would catch them before they reached help. As CIO, the police answered to him, and they patrolled the park roads.

Taljaard must have considered that. Via air was his best choice. That left Kagona's helo. It had armed guards, but Taljaard couldn't fly and his helo pilot was gone.

Kagona pulled the radio out again and located his pilot, who was helping guard the helo.

"Go to Brugman's office and stay there until I arrive," Kagona ordered. "Have one of Brugman's men escort you. Taljaard might try to force you to fly him out."

"I understand, sir, but we are short of men."

"Do what I say."

"Yes, sir."

One other possibility existed. Taljaard's *wife*. If she was the pilot who flew a light craft in from South Africa, could she also fly a military Alouette? Kagona's contact in South Africa had

246

given minimal details on the craft sent for the boy, although from its ease of landing anywhere and remaining unseen, he guessed it a bush plane, gyrocopter, or ultralight. Who had sent her to twist Taljaard to their needs, and who had trained her?

Damn, why didn't his contact answer his text?

As though bowing to his silent demand, Kagona's phone vibrated. A message brought sickening clarity. The redhead wasn't just any pilot. She was a helicopter test pilot.

One of Kagona's men in fatigues and bulletproof vest strolled a perimeter around the Alouette helicopter. Most modern countries had long removed this older flying machine from their military arsenals. Joni might need an act of God to get the thing airborne.

"From your look," Ian said, "I take it you haven't flown this model before."

"If I can get her started, I can get her in the air. A military aircraft will have an engine start checklist in the cockpit somewhere."

"Here. You'll need this." He handed her a mini flashlight with a blue-green lens. "There won't be time for much manual reading, though."

"You'll have to make time. I'll need a minute or two for orientation or I'll get us all killed."

"I still have a few tricks left up my sleeve. You'll have your time. Too bad Mike's not here. He flew this one back in the seventies. Said it's a fugly piece of machinery."

"Fugly?"

Ian pressed his lips against her ear. "Fucking ugly."

"Great." When pilots used such an endearing term, they were either talking looks or handling characteristics. She sure as hell hoped this chopper didn't fly like shit.

Ian knelt next to Sipho. "Do you think you can do what we talked about?"

Sipho nodded tentatively. If he remained silent, Ian's plan wouldn't succeed.

"Work your way to that palm on the edge of the grass. Stay behind the tree and well back of the brush under it. Wait to hear me make the first call. I want them to come to me, not you."

"*Yebo.*"

Ian gave him another hug then sent him on his way.

He turned to her. "You get as close as you can, but keep the helo between you and the lodge. Don't make a move until I take out the guard. I'm counting on Kagona's man not being familiar with the creatures of this area. If all hell breaks loose, you're on your own."

"I'd say that's a given. I'll give you the all clear before you move toward the helicopter. Watch the rotors."

"*Ziko ndaba.*"

She guessed that was the Zimbabwe version of no problem. The boys moved off as though they belonged to the night and shadows. For the first time, she listened to the jungle. The running river had a deep rush and gurgle, night birds cooed and called, and other sounds she couldn't recognize filled the air. Creatures as well as men hunted in the cool, moist darkness.

A funny chattering sounded close by on the edge of the clearing near the chopper. The guard paused and listened. The chattering came again. He turned in that direction, and the underbrush rustled not ten feet from where he stood. The guard lowered his rifle into firing position and moved toward the sound.

"Come out or I shoot." With no response, he moved even closer. Joni tensed. Did Ian have something better in mind than jumping out of the bushes to surprise the guy? A knife throw would be blocked by his vest, and if Ian missed, the guy would get a call off to Kagona.

The snuffling chatter came again, and underbrush rustled. The guard took another step. Joni poised to run. Every second counted.

Back toward the lodge, someone scuffed a stone with their shoe and sent it rolling. *Shit.* Company. She slid farther back into shadow. One of Brugman's men appeared.

The first guard glanced away to check out the new arrival. His mistake. He suddenly wiped an arm across his face and rubbed frantically at his eyes. He pawed at his clothing, dancing around as though standing in a pile of red ants.

"*Zorilla, zorilla,*" he screamed in his native tongue.

She puzzled at his words and actions until a foul, nasal-burning polecat smell reached her nose. Whew. Skunk.

Brugman's man smelled it, too, and made no attempt to approach. The guard threw down his AK-47 and ripped his vest and shirt off, using the cloth to wipe the sting from his eyes.

The half-naked guard grabbed his weapon and ran toward the second man, who moved away at the stench. They backed up along the walkway until Brugman's guy stopped and yelled at the guard in English. The gist—*zorillas* didn't roam this area.

Brugman's man headed toward the bushes, careful not to get close and end up smelling like his partner. He fired a shot into the brush near where the guard had stood.

Joni couldn't risk Sipho or Ian taking a bullet. They required a diversion. With Kagona's guy more or less neutralized and struggling to get his fouled fatigue pants off, she launched toward the chopper. The man wiped at his eyes and looked twice before pointing and yelling. Brugman's man swung his rifle around but never had a chance to fire.

Ian took him down in a shoulder tackle, knocking away his gun. That's all she had time to witness as she sprinted toward the Alouette. Kagona's pilot had not attached rotor tie-downs, expecting a brief visit. One less preflight item to do. Sheesh, who was she kidding? This baby wasn't going to *get* a preflight check.

She slid open the crew door facing Ian, then yanked open the pilot's door and climbed aboard. With every second counting, she stuck a headset around her neck and set to work.

Whatever vinyl had made up the seat cushions or covered metal parts of the craft had long ago deteriorated. A thick reed mat padded the seat. Her intuition screamed at her to bag this bucket of bolts for any other safer option. Yeah, right. There wasn't anything safer.

Pilots usually tucked the checklist between the seats or in some little nook or cranny on the craft. It was nowhere. Who trained these guys? She caught a glance of white behind the anti-torque pedals. Stretching out of the seat, she swept up a thin ringed booklet. English, the universal aviation language. Perfect.

She did a quick perusal of the cockpit, locating switches, gauges, and throttle. She'd started enough different turbine engines to assume the sequence didn't vary much. The NVGs had a gap that allowed her to look down and read using Ian's small light. A quick flip through the checklist pages brought up engine start.

Battery on. The gyros whined as they started up. Fuel valve and start switch on. The battery powered the starter, which turned the compressor, and the turbines started to move. So far so good. She slid a finger down the checklist for the minimum pressure required before she could move the throttle from cutoff to idle.

"Come on, come on, Fugly."

She shoved the throttle to idle, sending fuel to the engine. It ignited, and she watched the exhaust temperature go up. If it went too high, Fugly was going nowhere.

Her gaze flicked outside to Ian and the guard grappling on the ground, and then down the path toward the dock, where several men suddenly appeared—one of them Kagona.

They'd run out of time.

Ian must have heard the commotion, and with a last-ditch effort rolled, taking the man's arm with him. She didn't have to watch to know Ian snapped a bone. He launched to his feet. He still had to retrieve Sipho.

With the exhaust temp stabilized, she advanced the throttle to full rpm, and the whop of the blades changed to a higher-pitched whir as the turbines accelerated. The chopper strained with power, ready to lift off. A slight pull of the collective and the chopper blades lost their dangerous droop. The Alouette grew light on its wheels.

She signaled Ian all clear. He ran full out for the chopper with Sipho under an arm and his pack over a shoulder. Ian tossed Sipho on board but didn't climb in.

"Get in here, Taljaard."

A bullet whizzed past the door and pinged off the chopper.

Ian grabbed something from his pack. With a quick turn, he aimed a tiny box toward the shooter and pressed a switch. Smoke rose up along the trail.

"Sipho," she yelled back over her shoulder. "Grab a seat and get the harness on. We're in for a wild ride." She slipped a headset over her ears.

Ian dived into the chopper. "Go, go, go." He slid the crew door closed.

Not needing any further encouragement, she pulled the collective upward and they lifted off. Ian secured Sipho before grabbing a headset hooked by the door and popping it over the boy's ears. Then he snatched another pair for himself.

"Be prepared to follow orders, Taljaard. You're in my territory now."

He grinned. "Aye, aye, Captain."

"Really, Taljaard? That's the navy."

She heard him chuckle as smoke from a grenade below billowed and swirled. Kagona gaped up in disbelief as the downwash hit him and his men. Unable to help herself, she flipped him the finger. Whether he saw it or not, she didn't give a damn. They were away. Score one for the good guys.

Kagona strung a line of curses longer than his name as he ran for Brugman's office. He wanted that woman dead. Next time he had her in his hands, he'd watch the crocs devour her limb by limb until the last chunk of sickly pale flesh was gone. That'd remove that superior smile, proud chin, and foul mouth from her face.

His pilot, who had appeared at the commotion, leaped out of the way. The man strolled calmly after him into the office.

"I want them stopped." Kagona corralled his rage to think. "Find out if our air force has planes available to intercept them."

"They would not get here in time. But it's not a problem."

Kagona raised an eyebrow. "It's not?"

"Yes, sir. I know exactly where to find them." The pilot grinned, stretching out his moment of glory.

"Then perhaps you'd better explain before I stake you by the river."

The man swallowed hard, wiping the grin off his face. "Fumes, sir. The woman. The Alouette. They are flying on fumes. Remember how I suggested we land at Victoria Falls Airport before coming here, but we pressed on? The helo needs fuel. The only airport with fuel in range is Vic Falls."

Hope rushed through Kagona. The knowledge returned his power, his control over the situation. "Get in touch with the airport. I want the police, customs, military, everyone at the airport."

The pilot nodded and in minutes was connected by satellite phone to the airport. The conversation took much longer, as the people on the other end either were damn lazy or didn't understand the implications of their noncooperation. Kagona snatched the phone from his pilot's hand and straightened them out.

A haggard Brugman walked in as he finished.

"We'll take your boat." Kagona pointed to Brugman. "Where have you been?"

"Taljaard has wrecked my lodge. Smoke grenades and flash bangs everywhere made a mess and set a small fire. The main generators are toast. My storeroom has a gaping hole in the thatch." Black soot marred his light slacks and shirt. "And my control room..." The fool looked close to a breakdown.

"The blame isn't mine. You brought the boy here; it's where you chose to exchange him. You're coming with me, and I need your boat captain."

"He's with your men." Brugman acted sorely disappointed. His expected fortune had fled with his guest's transportation. "I'll get the keys."

Brugman punched a button on his desk, and a small key safe swung open near the door. Kagona sincerely hoped Brugman met his demise at the hands of one of his gadgets. Kagona pointed at his pilot and then said to Brugman, "Meet us at the boat."

As Kagona stepped outside the lodge, a boom and shudder shook the night air. It came from an explosive one of his men had planted on Taljaard's wretched boat. The pilot thought he had escaped. Fool. Not today. Not ever.

Enthusiasm filled Kagona. He slapped his own pilot on the back. "We have taken out one of those wishing to destroy our government. Get me to the airport. There we will get the rest."

CHAPTER TWENTY-FOUR

Joni relaxed once airborne, realizing how her breath had caught when bullets pinged against the chopper. "You nearly ate a few bullets. That little delay could have cost your life. We were okay."

"In case you didn't notice, fly girl, they were aiming at you, not me." Ian, who sat in a rear-facing front seat, poked her in the shoulder. "Kagona wants my hide alive."

His answer surprised her. Was she really more than a means of escape to Ian and a person to rescue Sipho? Did he truly care about her—the woman and pilot barely holding herself together? No matter how hard she fought against letting him past her defenses, he destroyed her attempts at every turn.

"Thanks," she managed.

"Seemed only fair. You saved my arse with the dash to the helo. Nice timing."

"Don't get overconfident. We might be above the jungle, but we're not out of it yet. We have to locate Mike."

A little fireball on the river below nearly blinded her. If it hadn't been for the automatic filter on the NVGs to protect vision when exposed to bright sources, she wouldn't be seeing anything at all. Seconds later the percussion shook the chopper.

"Bloody hell, that was bright," Ian commented as though it was an everyday occurrence.

Below fiery pieces died out in the water, as did hope of seeing her newest friend again. "My God. I told Mike to take the boat downriver to a good spot for pickup." Her words flowed faster as the implications of what happened struck. "Kagona must have rigged it with explosives after we arrived. Mike even said he thought something wasn't right with the boat. I should've helped him look it over."

"He made a proper show of going down with the ship, I'd say."

"You're sick, Taljaard. Your friend has just been blown to a thousand bits."

"The boat certainly was. I think Mike added something extra for pizzazz."

She looked over at him. "What the hell are you blabbering about?"

"Mad Mike has a nose for explosives. Elite Selous Scouts, like him and my dad, had more than one specialty during their fighting days, you know. When he wasn't flying, he was blowing things up. Come on, Joni, we expected Kagona or Brugman to find our boat. It kept a few men occupied while they set the charges." Ian pointed out the cockpit window. "Now, keep your eyes peeled for his signal."

A light flashed up from below in a series of blinks. Ian punched a fist into the air. "Oh, yeah, that's him, all right."

"Shit," Joni yelled.

Ian looked over at her. "Two seconds ago you were crying over the poor guy. Now he's shit."

"No, it's not Mike. We're out of fuel."

"That's a problem."

"A mission-ending one, I'd say." She started an approach to a hover. "How well do you swim?"

Ian hauled out his pack and grabbed his NVGs. "I think

we're fine. Kagona's pilot must have left enough to get to an airport. We'll make it to Vic Falls." He slipped the securing straps over his head and stared out the front bubble window. "Do we have enough to collect Mike or will we have to come back?"

Joni had no idea what shape the fuel gauge was in or if it even worked. Relying on Ian's logic, they had enough to pick Mike up, but whether they had enough to get to the airport afterward was another story. Coming back for him posed its own problem—the least of all was the entire Zimbabwean Air Force looking for their chopper.

"She's breathing fumes, but coming back isn't a good option, either. The worst that can happen is we'll all join Mike." Sure, she sugarcoated the truth, but why scare Sipho further. The kid's eyes were already wide orbs.

Mike sat on the flat basalt of a rock island in the braided Zambezi River. The falls lay farther downstream. She headed the chopper into the wind and allowed the craft to descend. She passed over Mike at a hundred feet, keeping him in sight and checking the terrain. No major obstacles.

"Get a tether on and slide open the crew door," she commanded Ian. "Lie on your belly and tell me what's under the chopper. You're my eyes for touchdown."

She made a right turn so Mike remained in sight at all times. With finesse of the controls, she hover-taxied to about six feet, where Mike, with the wind at his back, knelt at her one o'clock position.

"Talk to me, Ian."

He hung his head out the door. "Take her down. It's not perfectly level, but enough to get two wheels down."

She lowered the hover.

"Half a meter," Ian called. "Hole under right wheel."

They crept lower until she felt something touch.

"Nose and left wheel down," Ian confirmed.

Two wheels would do. "Go and keep your head down."

Ian released his tether and dropped out to help Mike. They locked arms over each other's shoulders and swept back up to the chopper. Ian boosted Mike in the crew door. Mike scooted across the floor to even the load as Ian climbed back on.

She lifted off and gained altitude while everyone settled in. A pool of water gathered around Mike, but he smiled and held up his leg. "We both made it."

"Strap in, boys. We have to refuel this girl before she croaks."

They grabbed seats on either side of Sipho, and Ian tossed Mike a headset. "Whew. Rescued by an angel. You're a lost man's dream." Mike launched into a blabbering of thanks. "Not bad flying, either, missy. You'd make a hell of a tour operator."

"No, thanks. I prefer getting shot at or eaten alive. It's easier." She pointed at the foot he still held in his hand. "You two are a hard act to part."

"Damn right. Breaking in a new prosthesis is ruddy hell. So where are we headed?"

Joni tapped the fuel gauge, noticing the peg had dropped farther than she thought possible. "Vic Falls Airport for fuel. She's breathing fumes." Any moment she expected the main rotor and engine rpm gauges to split, with one in the green and the other indicating zero. Maybe the chopper would vibrate and shudder first as a warning, or simply lose power.

"In this neck of the woods there isn't any other option," Mike said. "Vic Falls is closed, though. It'll take an act of God to get anyone out after dark. They'll hassle us to even turn on the landing lights."

"No choice but to give them a try. Know the frequency, by any chance?"

"In my sleep." He rattled it off, and she dialed it in. "Good luck on getting a reply at this hour."

Surprisingly she raised a live person and wasted no time telling them what she wanted. "This is Alouette 5124, requesting a landing to refuel." She gave a different call sign than the placard on the dash, just in case Kagona had radioed an alert for their craft.

"Vic Falls closed. Emergency landings only. Refueling impossible till morning."

"We have a sick child on board. Need to transport him to Bulawayo."

"No fuel available till morning. Over."

"Fifty-one twenty-four willing to pay double for fuel."

"Copy, Alouette 5124." The speaker paused for a few seconds. "Cleared to land. Further directions upon visual."

"Thanks. Fifty-one twenty-four out."

"Nice touch about the kid," Ian said.

Joni didn't feel quite as excited. Something bugged her about the exchange. After hours, African airports were notorious for charging exorbitant landing fees for everything, including flipping on a single light. Double didn't even come close to what some charged. The guy had made no effort to talk up the fee. Not even the smallest attempt. Nor did he spend ten minutes whining over the effort of calling in refueling guys.

"Mike, what do you think?" She looked over her shoulder.

"That was easy. Too easy."

"My thoughts exactly. Any bets Kagona knows we are coming? His pilot would have known the fuel situation."

"He has us over a fuel barrel, missy. Fuel is hard enough to come by, but in this part of Zimbabwe there is nowhere else. We could cross the border to Zambia, if you don't mind being thrown into a concrete prison for entering their airspace without permission, or handed back to Kagona. Of course, at least we'd be alive."

"Any other ideas, guys? We can't hover forever."

"Find somewhere safe to set down where we can hide the helo and make it out on foot," Ian suggested. "It won't be easy, but we'll survive."

With Ian for a guide, she knew they'd survive, but they didn't have the time. "Taz can't sit around for a week. Discovery is too likely."

"Not many choices," Mike chimed in. "Should we take a consensus?"

"Damn it. Kagona is probably celebrating." How the hell had she not noticed the fuel gauge when she started the chopper? Basic mistakes kill. Of course, if they hadn't cut a few corners, they'd be dead anyway. Some trade-off.

She rubbed at her chin in thought. The scent of fuel played on her skin. "Crap. The obvious is the hardest to see. We'll beat Kagona at his own game." She felt a wonderfully devious smile spreading onto her face. "There is one other place to get fuel. While Kagona masses forces at the airport, we'll be tanking up somewhere else."

As soon as she changed heading, Mike chuckled. "I follow you now. That cold river water must have slowed my thinking."

Joni had grown awfully fond of this codger, with his indestructible spirit that matched his name. At last she understood the true meaning of "Mad" Mike. She laughed. "Can you imagine Kagona racing to the airport to catch us there?"

Mike's lips thinned, like a child guilty of indulging in one too many cookies. "Don't imagine too hard, missy. He'll never make it. Not if he took Brugman's boat. It should have sprung a real big leak as soon as the throttle hit three-quarters power."

"He's in for a swim, eh? That'll buy us a little time." She made a fast trip with some guidance from Mike's local expertise to plant the Alouette as close to the disabled Jet Ranger as she could make it. She suspected the hotel employees were surprised by a night arrival on an unlit lawn. Hopefully Kagona hadn't thought to warn any of them about the chopper's theft.

"You want to take the pilot's seat, Mike, while Ian and I handle this?" Joni slid off her headset.

"Actually, missy, I think siphoning gas will take too long.

Once we don't arrive and one of Kagona's flunkies reports us landing, we'll be toast. But I've another idea. One that'll make this stop a whole lot quicker and us harder to catch."

He offered two words.

Ian laughed. With chagrin, Joni nodded, realizing she'd missed the obvious. "I can't use cold water as an excuse."

Ian slid open the crew door. "That's why we work as a team."

Mike tapped Ian on the shoulder. "Grab a bottle of water from the Ranger while missy's working? I'd like to clean up and work at getting this leg back on."

Joni unlatched her door and shoved it open with a foot. A solid, wholesome, good feeling engulfed her. She had a live kid in the backseat, a new friend behind her, and a man she couldn't get off her mind. Now, in the hours before daylight, all she had to figure was how to fit them into her life.

CHAPTER TWENTY-FIVE

The good-bye to Mad Mike had been brief by necessity after Joni landed in a clearing within a mile's hiking distance of Taz. Everyone but Mike off-loaded, and he, with his prosthesis cleansed well enough to wear, had taken the pilot's seat. Before they had reached cover, he was gone into the night with his expensive goggles, nine millimeter, and a damn good story to tell. She hadn't bothered to ask where he was headed, but she suspected the Jet Ranger would be tucked away before daylight.

The turnaround back at Victoria Falls had been quick. She simply borrowed the larger battery from the Alouette and placed it in the nose of the Jet Ranger. It required a little finagling to fit, but functionality was all they required. Taking the swifter Ranger also guaranteed Kagona couldn't catch them if he reached the Alouette and was able to refuel and find a battery for it.

"Are you going to stare at Sipho all night?" Ian asked.

Joni stood inside an abandoned barn once used for milking before it was converted into storage and stalls for family animals. The farmhouse had burned and the land had since been claimed for a collective farm that had yet to materialize. They huddled in the loft, where Sipho lay curled up on a pile of soft hay, with Ian's dark cap tugged down over his ears, a light jacket on his arms, shoes laced up tight, and his little pack as a pillow.

"I thought he'd never let go of you on the flight here." Sipho's chest rose and fell as he slept. Ian had dug out a glow stick, which Sipho clutched, highlighting a fragile peacefulness on his face. Six hours away at sunrise lay one last step to his safety.

Ian touched her shoulders from behind and stepped close enough his heat warmed her body. "You're tenser about the morning than he is." His fingers massaged the taut muscles of her neck and shoulders. For a man capable of rendering someone unconscious in seconds, he had surprisingly gentle hands, ones she didn't want to stop.

"Hard not to be. I've seen what he'll face if we're caught. The military will be on the lookout for Taz."

"Is that the only reason? Your first reaction when I touched you was to tense even tighter."

"I was thinking about Sipho," she lied, not about to admit that today's adrenaline rush heightened her intimate feelings toward Ian. "He told me this camo paint made me look like a *sangoma*."

"Consider it a badge of honor. I've something that will help." He reached into an outside pocket on Sipho's pillow and pulled out a thin packet of baby wipes.

"Why, Taljaard, is there something you're not telling me?"

His chuckle gave the vestige of security where little existed. "It won't clean the paint out of your pores, but it's a start." He then bent something else in his hand and took her wrist. "I saw a kerosene lamp near the barn door. If it has any fuel in it, we can use that to clean up."

"On my face?" A soft light came from a glow stick he wrapped around her wrist.

"Might take off a layer of skin, but it works well."

"Thanks, I'll skip the carcinogens and wait for cold cream or a hot shower and soap."

He lifted her glowing wrist toward her face. "I think you're right beautiful, no matter what's on your face."

She swallowed hard. From his picture on the briefing room wall to his first doubtful look upon her arrival, he had sparked something inside her. This whole nightmare, or deadly adventure, or whatever a crazed person would call it, made her want to fly home and hide under the bedcovers, hoping the world would go away. No more people being eaten alive or beaten or shot. But deep inside, it made her sad to know these new friends, ones who would literally give their lives for each other, would be lost.

He pulled a damp wipe from the packet. She rested her hand on his shoulder so the glow highlighted both their faces. With a firm but gentle pressure he wiped along her cheekbone.

"It's cool." A shiver shook her body. "Lack of sleep must be catching up with me."

"Nights here can be chilly. Put this back on."

He tugged the cap she'd worn on the mission onto her head. He used the excuse to circle an arm over her shoulder to support her head while he cleaned. The heat from his body had a disarming effect. Each touch brought into focus how close he hovered, how caringly his large hands caressed, and how badly she wanted more.

He snagged another clean cloth. His eyes meticulously followed his efforts as though recording each groove of her features for posterity. "I've never worked with anyone like you before."

"You mean a woman?"

"Not exactly." He struggled for words. "They were right in sending you."

"I was simply in the right place at the right time."

"And what place was that?"

"Right where your request found me, flight testing Taz in

South Africa. I was selected simply because I'm small enough to pilot the plane and still have room for Sipho."

"Lemmon isn't a fool. You're a former military special ops pilot. I've seen what they can do. You've natural instincts."

"Ones that almost got me killed."

"You should trust them more." His face hovered close. "I saw you in the air. It's the place you feel strongest. With more training, that could carry over to other arenas."

"I've told you, Ian, I'm just—"

"—a pilot." The kiss came quick and light, one meant as a mere introduction to what they both wanted to follow. Much too soon, he drew back. "One with convictions and the backbone to see them through."

He seemed to sense his move left her off balance. His hand slipped from behind her head and instead held her chin steady while the other hand wiped firmly along it. After dirtying another cloth, he placed a soft kiss on her forehead.

"Good to go." He handed her the packet. "Your turn."

She hesitated. He knew what contact with the face that had given her strength through the day would do to her. She had to keep control while she still had any and get answers about him that had been dangling out of reach.

That effort proved hard. The cloth chilled Joni's fingers, as she stroked above his brow. Yet heat tingled through her fingertips from his warm skin beneath. Ian purposefully played with her, trailing his hands from her waist up her sides and allowing his thumbs to stroke over her breasts.

Joni had hardly cleaned the bridge of his nose before he fingered a strand of hair and used it to tickle her cheek. She did her best to ignore him. Soon dark camo grease covered the wipe and she let it drop. He caught it and slipped it in into a pocket. *Leave no trace behind.*

While she dug for another cloth, his eyes studied her, invited her closer.

"If you're a man of your word, Ian, I believe we have some unfinished business." She moved in as though deciding where to clean next.

He secured a possessive hold and leaned in for another kiss. "I thought you'd never remind me."

She angled her head at the last moment. Her mouth slipped to his ear, and she brushed her lips against him. "What was missing in your bio?"

"Crikey, are you extorting information from me?"

"You promised to tell me more about your background after we secured Sipho. Besides, it's only extortion if I have something you want."

"What if I want you?"

Those words fired a sensuous jolt through her body. Did he? She rubbed the wipe along his jawline. The rough stubble pricked her fingertips and inched up her heart rate. It had been a good while, but she recognized the signs of slipping control. The intensity in his gaze said he focused fully on her every touch. She lifted her fingers from his jaw and, unable to stop, traced one along his lower lip.

He wrapped his hand around hers. "If you start this fire, you'd better know how to put it out." The muscles of his arm became taut about her waist, drawing her against him.

"I'll risk it." A perilous decision, as her body had already moved into heated territory.

He dropped the baby wipes back on the pack beneath Sipho. "Sounds like we need privacy for negotiations."

He prodded her down the loft ladder to the colder, haunting barn below. Old hay and dirt crunched under her feet on the barn floor, and musty smells filled the air. Shadows loomed everywhere, and she imagined a threat hidden in every one. She moved closer to Ian, who seemed at home in the darkness. They stopped at a wheeled metal object that looked like the back of a pickup loaded with hay.

"Is this a vehicle?" She held up her glowing wrist, but the light didn't cut deep enough into the shadows.

"It's a farm wagon. Hasn't moved in years. The current owners have no tractor to pull it." He dropped the tailgate. "This isn't a perfect conference room, but it will have to do. Have a seat."

Ian sounded so businesslike she feared the momentum she'd gained in the loft had been lost. She stood her ground, merely leaning a hip against the wagon. He stayed close, his gaze measuring up her demeanor.

"So where do we start?" She crossed her arms, ready to play his game. "You promised to fill me in on this mission and your background."

"Not until I have a sign of your good faith." He unfastened his holster and placed it on the wagon. His heavy weapon belt followed. Even without it, he didn't appear vulnerable.

"You aren't serious."

"Totally." He stood before her with arms relaxed at his side and an expectant look on his face.

"Lemmon should have mentioned what a cocky bastard you are."

She sidled into his space. With chin up, she rose on her toes and brought her mouth near his. He lowered his head, but she leaned back. With arms crossed, she grabbed the edge of the black shirt she'd worn over her tattered blouse and pulled it over her head. Her shoulder screamed in pain, but she refused to let it interfere. She tossed the shirt on the wagon.

He raised his brows. "Interesting token. Are you getting hot or demonstrating the first step in more to come?"

Her fingers latched onto his shirt and she tugged herself closer. "Start talking and you'll find out."

A hint of humor flickered in his eyes. "Since you won't take a seat, I will." He parked on the tailgate, swinging his legs off the end. That boyish charm he surely used to seduce women was in full force...and it was working.

"I told you my dad was a Selous Scout. After his days with them ended, he took over the family ranch. He tried to continue the success my grandfather had built but never had ranching in his blood. When the time came for me to attend a good university, we were short on money."

She wrapped her arms tight to chase away the chill and faced him, settling against his leg. "So how did you pay for it?"

"I went to Britain to work for a while. Eventually we had enough for me to go to school."

"That doesn't account for the way I saw you move, or plan, or execute everything you accomplished today."

He trailed a piece of hay up her arm to her exposed neck. "Like how I took you down near the stone bench."

"That qualifies."

"Krav maga. Spent years getting a certificate. The kissing part came from a different kind of practice. Did Lemmon include the part about my British girlfriend here in Zimbabwe?"

"It was mentioned, but—"

"That was one of the bigger mistakes in my life. She lasted a month in country. The unstable political climate did her in."

"It nearly did me in, too. But you're evading the information you promised."

"She isn't in the same league as you."

"You aren't going to tell me, are you?"

He raised innocent brows. "I've given you something you didn't know before." He gently grasped her chin and brushed his thumb along her cheek as though wiping away a remnant of camo grease. "It's time for another sign of good faith. Shall I help you with the next piece of clothing?"

His fingers popped open the first button on her tattered shirt.

Things progressed too fast considering the number of questions yet to come. She caught his hand. "The buttons are all you get...for now."

He had them undone before she finished the last word.

Minus the overshirt and with her buttons open, she shivered in the night air.

"Come here. I'll keep you warm." He twisted her around to stand with her back resting against his chest. He wrapped his arms around to her stomach and brought his legs tight against her hips. "How's that?" He tucked his head over her left shoulder.

"Warmer. And if it's meant to soften me to your excuses for not talking, it's working."

"That's what I like about you." He lifted her cap off and tossed it behind him. "You're real." He kissed her hair. "Not afraid to be honest about what's inside."

An odd sadness encompassed her heart. The time spent with Ian since she'd landed had turned those first photo impressions of the curly-headed blond into an emotional collage. Her defenses melted with his smug smiles, the soothing tenor of his voice, the way his arms wrapped her in hope, and his loving concern for a child.

"That's the problem." Her tone sounded more pensive than she intended. "I'm not real. Hours from now I will disappear. All of this will be a vivid but brief memory."

"You're real to me. Right now. Right here." He pushed aside her hair and brushed his lips along her neck. "Way too real."

She focused on his warm, moist touch as he trailed his mouth up behind her ear. Goose bumps rose along her arms. If he continued touching her for much longer, her ability to keep him distant would disintegrate.

"This isn't a game, Ian. I can protect Sipho and myself better if I know the truth about you and this mission."

Grasping her shoulders, he rotated her to face him. "Trust me, I plan to answer your questions and more."

"But will you tell me the truth?"

"The truth is...I don't want to talk." He lightly kissed her brows. Then the tip of her nose. "Not now." His hands slid under her loose shirt while he planted more kisses. "Later. Everything. I promise."

By the time his mouth touched hers, the desire to maintain control had fled. Hell, what control? She didn't have any, only a yearning to wallow in intimate sensations and forget the horrors of the world. Pent-up emotions from the day spilled forth from Ian, and she found it impossible not to match his intensity.

His tongue swept across her teeth, seeking entry. Aware of what her mind and soul craved and unable to deny him the same, she opened the kiss to him. He delved deeper while his palms burned their way along her back, finding the edge of her bra and following the cloth like a map to a destination. In a breath, he released it, removing another layer of distance she had kept from intimacy.

His thumbs slipped to the tips of her breasts, teasing them to harden beneath his touch. She fisted his shirt. Any questions remaining in her head slipped away to a singular need for him.

He released her and slid off the wagon. In a quick pull, Ian removed his shirt and spread it on the tailgate. Using his hands to speak for him, he backed her up against the wagon and ran them the length of her body, claiming it and assuring him she truly was his, at least for the moment. He slipped her shirt off her shoulders and took the bra with it. He added them to the growing pile on the hay before entangling a hand in her hair and engulfing her mouth again.

The cautions, the warnings, heck, anything that sought to slow or stop what was to come, she shoved to the back of her mind. She'd forgotten how smooth and hot a man's skin could feel. His chest pressed against hers, taking away the night chill and wrapping her in warmth. Her mind switched from thinking to experiencing.

His intensity mingled with hers. So intent on his kisses, it took a moment to notice he had cupped her bottom and straightened, lifting her into his arms.

His ease at holding her offered assurance he had the ability to handle the complications between them. The soft glow on his face lit hungry eyes. At her scrutiny, he gave a wicked smile and set her on his shirt, leaving her legs dangling.

He worked his hands down her leg. Off came a boot and sock. He followed with one of his. A quick repeat on the other side had both their feet bare. He stroked the bottom of her foot, and a surprising growl released from her throat. She wanted more...now.

Encouraged by her reaction, Ian drew her into a slower, more languorous kiss, one that crushed critical thought and focused on pleasure. He reached beneath the band of her black pants and unsnapped her cargo shorts. The zipper gave a soft burr. With a slight lift of her bottom, he tugged everything off in one sweep. Another layer of clothes fell to the hay behind them.

He stroked her legs, bringing back memories of the kraal cleansing. Wanting to see more of him, she reached for his pants, but he backed out of reach.

"I need to use a condom." His breath huffed, and he gave her an odd look.

Frustration?

Her emotions, raw from the night's perilous adventures, understood the gamble they took, focusing on each other without protection. "I can still touch you, Ian."

"It's not that. *You* matter to me. What you *think* matters to me. Bloody hell. I don't want you thinking I screw every girl I meet."

Confused, she didn't understand until he pulled a condom out of a pocket. She couldn't hold back a soft laugh. This tough, assured, in-control guy worried about her health *and* her feelings. It rather brought him down to earth.

"I was afraid you'd react that way."

"A tool of the spy trade?" she teased.

He positioned his hands onto the wagon on each side of her

thighs, leaving them nearly nose to nose. "No, I don't carry condoms on missions, although they're useful for survival."

She raised questioning brows.

"They're perfect for carrying water." He straightened and raked his fingers through his hair. "Remember how Mike and I spoke privately before he left?"

"Must have been an interesting conversation. Remind me to thank him." A chill whispered along her bare skin, and she shivered. "But right now, I'm cold. Either dress me or warm me up." She circled a finger around his belly button before hooking it on his waistband. "Your choice."

To motivate the decision, she unzipped his outerwear and went for the khaki shorts.

His jaw clenched at the contact. "It appears negotiations have been postponed." He ripped open the condom, and she slipped it from his hand. Her legs tightened around his hips, bringing him closer.

He hovered over her. "You're a ruddy tease." His kisses became rougher, full of need and anticipation.

The decisive soldier in him moved into a full offensive. This was not to be a slow, methodical seduction, but a swift, memorable maneuver to bring his opponent down and leave her begging for mercy. His fingers aimed for the territory with the highest return for his efforts.

Her breathing degraded to quick pants, growing more rapid as the sensations heightened. The devil grinned with that rugged, dimpled look meant to disarm even the wildest creature.

"Damn it, Ian." Unable to withstand the sensory assault, her hands strafed up his arms and her nails dug into his shoulder. "You knew from our first meeting I couldn't resist you." Her mind and body rushed toward overload. Her head dropped back under his onslaught.

He slid her hips toward the edge of the wagon and thrust inside, carrying her over a peak in a wash of sensations. His movements became more and more driven as crest after crest shook her body. He finally held himself deep, groaning as he buried his face into her hair.

She held on, not wanting the moment to pass. Every nerve cell she owned had been dragged through heaven and back again. With sensations still tingling in her body, she kissed his forehead and let what little fingernails she had lightly scrape up and down his back. Strength she had witnessed in Ian's every move lay exposed to her touch.

He lifted his head and placed proprietary kisses across her cheeks and neck.

Elation and warmth swept away worries of what might be and let the here and now take over. Too many years had passed without a man in her bed. A yearning for more intimacy filled her heart—intimacy with this man.

She sighed and released him, lazily stretching back on the clothes.

He fastened his pants back on before settling beside her on the wagon. "Ruddy hell, they don't make pilots like they used to." With a satisfied sigh, he hung a leg atop hers and hooked an arm at her waist. "How many more times do you think we can repeat this before sunrise?"

Joni's warmth burned under his touch. In the soft light, she looked like a golden goddess—one who understood his solitude, his lifestyle, and yet his needs.

Her laugh carried contentment. "Haven't you ever heard of crew rest?"

"Sure. You sleep and I'll watch. Don't blame me if I touch a little, too." He wanted much more but would do anything to make sure she and Sipho were ready to fly.

"There's still time for play. I won't be gone until morning."

"If I had my druthers, I'd never let you go." He propped his head up on a hand and kissed her on the nose. "A day ago I couldn't imagine I'd be in this position."

"By my side or on top?" She swept a strand of his hair back and grinned.

"Both." He couldn't help but taste those teasing lips again and pulled her body against his in a move to possess an impossible dream. When he lifted his mouth from hers, yet hovered near, a profound sadness filled him. A future together *was* an impossible dream. "Our line of business has no guarantees, Joni. No matter how badly we wish for something."

"Wishes take time to come true. I'm patient."

"Waiting isn't the problem. It's surviving. I've no idea what it will take to free Sipho's father."

"I can be at your side—"

He pressed a finger to her lips. "Wouldn't Kagona love that? No, it is dangerous enough for me here. I'm not risking you again. Lemmon has promised help, and I'll need it. You've another mission to complete."

The reminder of the flight to get Sipho to South Africa replaced the contentment on her features with a frown. She shivered. He rubbed a hand down an arm to warm it, but goose bumps rose under his fingers.

"Damn." He sat up and climbed off the wagon. While the heat of the day kept the loft warmer at night, the lower barn remained chilly. He scooted up the loft ladder and snatched the extra pack that had his shirt he'd removed in the boat.

In mere seconds, he returned to her and handed over the shirt. "Put this on."

She slipped the shirt up one arm but struggled to get it onto her other. When her glowing hand reached up to tug up the shirt, he saw a large, multicolored bruise on the shoulder without the cut.

"Why didn't you tell me you had another injury?"

"Because there was nothing we could do about it. And honestly, I had a few distractions from the pain."

273

A surprising sense of protection flushed through him. "Did Kagona do this?"

"Brugman. When I went flying off his boat."

The fact he hadn't noticed earlier struck him hard. Nor had he even checked into her well-being. He'd just assumed. Yeah, he'd been assuming a lot of things his entire life when it came to relationships.

Joni wasn't one of the guys, she was...hell, he truly couldn't answer. What did she mean to him, and Christ, what kind of lover was he?

"Mike said you'd been shocked and knocked out. But I'd no idea about this."

"I tangoed with a pole on Brugman's deck. Can't remember much about the dance."

He took a closer look. "You're lucky it's not broken."

"It's starting to stiffen. I'm hoping it doesn't swell up and impede the flight."

Ian dug out a small paper square from his pack. He ripped it open and dumped two pills onto Joni's palm. "These will help."

She downed the pills without question, showing the trust built between them. She also didn't ask for water—the little they had left was for Sipho. Her concern for the kid meant the world to him.

Before first light he'd make rounds and refresh their supply. For now, he cuddled back by her side, putting the nearly empty pack under her head and positioning her body so her little arse snuggled up against his groin.

She nestled her head partially on the pack and his arm. His other one draped over her waist, and he cupped his hand under a soft breast. Hell, he wanted more nights holding a woman this way...a woman who mattered to him. He stroked the wild mess of her hair, made that way by the day's trauma and the last hour. He didn't care. It rather added to her allure.

Together they seemed so right. Damn that reality had to have its say. They both had committed lives to important purposes, and neither at this point in time could give them up.

"Maybe I should get dressed," Joni whispered and snuggled a leg along his.

He gently kissed the top of her head. "I'll keep you warm for a while."

"Are you sure we're safe here?"

"I'm never sure, but always prepared. Did you hear those bird calls earlier?"

"No."

"My two friends who refueled Taz are keeping an eye on us. They'll signal if a threat approaches." And they had the ability to do a hell of a lot more.

"Your friends will suspect you're distracted."

Of that, he'd no doubt. He'd hear about this for years to come and didn't give a damn. "A good soldier can handle more than one distraction at a time."

Joni wiggled onto her back to better see him. "Exactly what was your job in Britain, Ian? You said you'd seen what spec ops pilots can do."

"Bloody hell, woman, I have to watch myself around you."

She traced a finger along his temple. "Someone besides your father trained you. You move in this world with your mind attuned to survive. I saw it in our encounter with the thugs at the roadblock. You took them out with swiftness I've seen in the Special Forces I used to fly. It took mojo to take on Brugman's lodge without a full squad, and yet you did it successfully."

"I didn't do that alone."

"You know what I mean. Your gun and knife are within reach behind you on this wagon."

"Sex toys?"

She raised a knee and slowly slid it up his inner thigh, closing in on a sensitive target. "British SAS?"

His background wasn't necessarily a secret, but he preferred to keep it under wraps here in Zimbabwe. Ian stopped her knee and nuzzled into the unbuttoned shirt. His tongue tickled her breast. Feeling her nipple harden beneath his touch ratcheted up his drive to take her again, this time like a damn lover and not some deprived soldier.

Joni pushed his shoulders away. "Distraction won't work."

"I'll try harder."

"Where did you work in Britain?"

He drew in a long breath and propped his head on his hand. "I applied for training in the British Special Air Service and used my time with them to help pay for my studies."

"I've heard they're hard to get into. Like our American Special Forces."

"We're a step up from them."

"I can see you're impartial."

"Once it's in your blood, you never forget." He stroked her delicate chin. "The training has come in useful, and once in a while people ask me for favors."

"Is that how you connected with Lemmon and his friends?"

He merely raised his brows.

"You aren't going to answer that, are you?"

Hell, no. "You said you had other questions."

Apparently resigned to not getting answers to every question, she moved on. "Exactly how did you get Sipho out?"

He played with the buttons on his shirt wrapped about her. What he wanted lay beneath the material. "Remember what I told you about Brugman's bottom line being money?"

"Yes, and from what I saw today, I think you need to reevaluate that particular belief of yours. He fed one of his men to the crocs."

Ian picked a piece of hay from her hair. "You should know by now, life is unpredictable. Kagona likely forced his hand on the killing. Too bad, really—Brugman lost his most loyal man."

"What do you mean?"

"My insider in Brugman's camp was pretty sure the transmissions to me had been detected. He hid his radio in the trunk of the most ruthless guy in Brugman's camp."

"How did you find someone willing to work at his lodge?"

"I didn't. Last time I took on Brugman, Mike and I had more time. We kept tabs on his employees, discovered the one most likely to help for a healthy sum. My insider found it quite productive to have more than one source of income. There was cash, small bills, in the pack you carried. It does wonders for cooperation. I knew exactly how to evade the surveillance and where to find Sipho."

"What about the anklet on Sipho? Brugman claimed he was the only one with the code."

"He wasn't. Sipho saw him open it to put it on his leg. Never underestimate a kid. Even so, I would have figured a way to get it off."

"How did you get in the storeroom without the security cameras picking you up? They caught Mike near the kitchen."

"Sipho wasn't in the kitchen storeroom. That's why Mike created a diversion there. Sipho was in the control room with Brugman, not twenty feet from you, or at least a closet inside it. Once Brugman left to see what was happening to his precious lodge, only a technician was left behind. That room was the one place without monitoring, because they rerigged the camera to watch Sipho."

"I saw live video of him sitting on those plastic milk crates."

"The riskiest moment was when I entered the room. He couldn't move or it would alert Brugman or Kagona before I was ready. I signaled Sipho to freeze. He almost succeeded, but couldn't wipe that smile off his face."

"What did you expect?" She stroked a tantalizing finger down his bare chest, reminding him the night was swiftly passing. "The cavalry had ridden to the rescue."

He took her touch as an open invitation and leaned down for

a kiss. She pushed him back. "Oh, no. Tell me more. How did you get him out without being caught on the camera monitoring Sipho?"

"I copied the last minute of video and killed the time stamp. When we blew the power system, he and I walked out in the dark. When the emergency generator cut back on, it showed Sipho all nice and secure—which he was, with me."

"What about the person manning the surveillance equipment?"

"He took a nice long nap. Didn't have the time to warn anybody." He lied, but once again the less truth she knew, the safer for all involved.

Joni looked up at him, curiosity reflecting in the light of her glowing band. "Lemmon is working with you for a reason. It's more than bank accounts or corrupt leaders. A newspaper journalist could cover that. What is so important, Ian? And don't give me my boss's line that it's because Mukono is a friend of the South African president. What did Kagona expect to learn from Sipho?"

"Whether he knows about a particular account…a very healthy account labeled *Chitima*. One his father asked me to trace." He absently stroked along her arm, knowing this knowledge would not sit well. "Zimbabwe is making odd transactions via the account. At least one includes the Chinese."

"They've struck deals with China before. They trade for weapons and infrastructure. Why is this so different?"

"No one is sure. The places the payouts are going from this account are unusual."

"Payouts? What are they buying?"

"I think that's what has people puzzled."

"Do you think that's what Mukono was going to tell you about?"

Her question had run through his mind a thousand times the last week. "I'm not sure. Something about these account transactions is different. That's what I'm investigating for Lemmon."

"Whatever it is must have dire implications. No wonder Kagona wants Sipho so bad. I can also see why the South Africans, as the bordering country, are concerned. Have you asked Sipho about this?"

"Indirectly, but I got nothing. It could be because I'm not asking the right questions. He memorizes numbers and names. Mukono only gave me one account related to Chitima. There could be more...or even other information he withheld until he had an idea of what this was about. Honestly, that's what I'd do if I were him. Even I'm not sure how much I can trust Lemmon."

Joni laid an arm across her forehead. "What about you, Ian? Kagona will turn up the heat. He'll expect you to come after Sipho's father. Will Kagona kill him before then?"

"Kagona won't kill Mukono outright because of his high-profile job. Kagona may renew his efforts to get information out of him, but exact details will be hard to obtain, especially when we slip his father the information that Sipho is safe."

"Do you know where he is?"

"No, and that's another problem I have to work out."

"Will they put him on trial?"

"We can't predict how the government will proceed with his supposed crimes against the government and the death of his wife. Until they get their hands on Sipho, all that is uncertain. Right now, my focus is getting Sipho to safety."

"So what will Sipho do until his dad is free? Does he have relatives in South Africa? Can they keep him safe?"

"No relatives outside Zimbabwe. Lemmon said he'd see to his care and safety. I have some thoughts on the subject, but the details can wait till morning, when we plan for your flight out."

Joni sat up. "You can't trust him to Lemmon. Sipho's been traumatized enough. He can stay with me."

His heart swelled at her offer. "That's not possible."

"Why not? He feels safe with me, and damn it, Ian, he's a good kid."

"You can't be with him all the time, and I don't want him going back to the project office."

"I'll take time off. Take Sipho wherever you say."

Ian hesitated, not wanting to risk her more but realizing none of them would be safe without her help. "I have a secure place for Sipho. That's the easy part. What's not so easy for me is sending you back to the project office. Someone in that office could be a traitor, and Lemmon will use you to find out who."

"Christ, Ian. We can't be sure. I think you're overanalyzing this situation. Besides, I'm not a spy. He knows I've no idea how to do clandestine stuff."

"That's what bothers me, but he sent you here and I'd rather you work with Lemmon on the inside than be ignorant of what's happening around you."

Joni lay back with her hands covering her face. Her takeoff in the morning would put one threat behind and lead her into another. She dropped her arms and looked up at him. "My part is easy." She smiled weakly. "Yours? I'm not so sure. When all this is over, will you stay in Zimbabwe?"

He had asked that question every day, and his heart had refused to answer. He wasn't a man who liked to give up. "My ranch is gone. The chances of getting it back are nonexistent. I could move on, like many people, but I make a difference here. I'm a damn good wildlife manager, among other things." He skimmed a hand along her outer thigh. "There are enough people and agencies trying to save what's left of the animals here to hire me under the table."

Joni shifted to face him. He hated to admit when she flew away tomorrow his life would feel empty. Times and lives changed. Maybe he had to consider changing his.

"Even if I succeed or fail in freeing Mukono, living in Zimbabwe will become more difficult. Kagona did his best to ignore me before, when I proved to be a small irritant. Now that I pose a direct threat to him, to the government, he'll be looking for me. The effectiveness of what I can achieve here will be limited." Hell, he didn't want to think about the future at this moment. His mind and body craved one thing and one thing alone…Joni.

"I'm sorry, Ian. I know how you love this beautiful country." Empathy sounded in her voice.

"It's not as beautiful as the woman before me." He touched her hair, thick and wild. "Don't forget me. I want you in my life. When this is over, I'll find you, no matter where you try to hide."

"Hide? And miss more times like this, cuddled in itchy hay in a drafty barn loving the man of my dreams?" She reached up and touched his temple, then let her fingers trail down his chest toward his groin.

He held back a groan as his body grew hard in response.

"Just in case I don't have a chance in the morning"—she walked her fingers back up under his chin and hooked her hand around the back of his neck—"I want to say good-bye now."

Ian rolled her on top of him, shifting her legs to straddle him. His shirt covering her fell open, exposing the pale shimmer of a body—one he planned to devour again. "I don't believe in saying good-bye."

"Okay, then don't talk. Just warm me up again."

He could handle that request. He swept his arms around her torso and tugged her down.

She brought her face inches from his and smiled at his eagerness to rekindle body heat. "It's been a pleasure working with you."

"My thoughts exactly." He touched his lips to hers.

CHAPTER TWENTY-SIX

Ian sat in barn shadows, wrapped in obscurity. Years of training, in keeping attuned to his surroundings for danger, prevented sleep. Security for his friends came first. The artificial peace of the darkness allowed him to evaluate what had happened and what was yet to come.

Alone, he could easily fade away, blend into the night or his surroundings, live life on the move and risk danger at every encounter. But now his vulnerabilities lay exposed—the young boy in the loft above, and the woman nearby, sleeping in the wagon.

His empty life had started to fill, and it brought him out into the open. No longer could he close his heart and expect to remain untouched. His parents had fled his life, but Nomathemba had enriched his view of this land, given him something worthy to save, and included him in her family. Her friendship had softened his loneliness, and now her death ripped open the void.

Sipho had given Ian unconditional trust, yet his poor safeguards had allowed his worst enemy to kidnap and traumatize the boy. Now Sipho's safety lay in removing him from Ian's life. The void only grew.

Ian ran his fingers along the fold of a small paper in his

hand. The image of Joni and Sipho flying away brought relief and yet greater emptiness. She filled a part of him he hadn't noticed was hollow. In all honesty, he dreamed of a family, one impossible with the life he led. He understood the terrifying position Mukono faced if his child were captured, especially with Nomathemba already snatched from his life.

A soft vibration tickled his wrist. Time to wake Joni and head to Taz. With Kagona on the watch, he wanted her in the air at sunrise.

He listened to her soft breathing. "Joni." He gently stroked along her forehead. "Wake up."

Groggily she opened her eyes. "Time to head out?"

"Yep, about an hour to sunrise."

Joni, dressed again in her clothes plus the black outerwear, sat up and rubbed at her bruised shoulder, suffering from sleeping on the hard surface. The glow stick at her wrist had nearly died. "How much time do we have?"

"Enough."

She patted around the wagon and dug into the hay, looking for something.

"Whatever is missing can't be important."

"I couldn't find my underwear earlier."

"Forget it. You can live without them."

She shifted aside more hay. "You said *leave no trace behind.*"

"I'll double-check the barn later. Wouldn't do for someone to come across black panties with a lacy edge."

"Ian." The sexy suspicion dripping from his name brought the night back in vivid, glorious detail.

"Toying with you is rather refreshing." He leaned in and kissed the top of her head. "I noticed when I took them off."

"Right." She gathered his pack that had been under her head and shifted it to the end of the tailgate. She scooted to the edge as he set her hiking shoes on the wagon. While she worked them on, he unfolded the piece of paper in his hand and spread it open.

"We've a little flight planning to do." He shone a penlight on the paper. "I realize your flight map is in Taz, so this sketch will have to do. Before Mike left, I asked him to arrange for fuel at a strip in South Africa near the Botswana border. It should be enough to get you all the way to Hoedspruit. I wrote down the GPS coordinate and suggest you memorize it." He highlighted the place on his crude map of southern Africa.

Joni studied it. "Botswana may be the shortest path out of Zim, but it would seem smarter to refuel at a strip northwest of Krueger National Park. It's closer to home. Obergen could easily drive fuel up there by the time I land."

Aviation was Joni's turf, but he had other reasons for his choice. "This flight path is part of my plan for Sipho's safety."

"I'm listening."

"You remember the uncle I mentioned?"

"Yes, in South Africa, I believe."

Ian explained in detail, and her doubtful frown softened. He dug a phone out of his pocket and placed it in her hand. "I suggest you report in once airborne and well away from here."

It took her a moment to realize it wasn't the phone Lemmon had given her. "Where's mine?"

He shifted uncomfortably on his feet but told the truth. "Disassembled and buried."

"What?" She pocketed the phone and returned to tying her shoes. "Why did you do that?"

"So no one could track us."

"Damn it, Ian." She pulled the laces into a tight bow. Very tight. "That was an encrypted satellite phone. It's unlikely anyone could intercept the signal, let alone decode it. And after losing Sipho, I wouldn't have made any more calls until aloft." Irritation bubbled under her words.

"I couldn't risk someone might have bugged it."

"So when you disassembled it, did you find a bug?" She finished with the other shoe and hopped off the wagon.

"No, but I had no way of telling if the software had been hacked. Nonetheless, evidence points to a leak in your project office. Someone is feeding Kagona information, and, at that time, your name was on the list."

She stepped back. "You didn't really believe that, did you?"

"Logic said I couldn't trust you." He moved closer, but she angled away.

"Logic? Aw, hell, I'm in trouble now."

"Joni, what I do is no game. When it comes to people I care about, I don't trust anyone." Christ, could he screw this up any worse? "My suspicions didn't start when you showed me the phone. They began back at Lubbe's farmhouse. I overheard Kagona ordering his men to bring me and the pilot in alive. He wanted the location of your plane. That was within hours of your arrival and long before we ever got near the kraal, where you made your call. How did he know?"

She tugged hay from her hair. "I was spotted by another plane, remember?"

The defeat in her tone showed her belief in him had taken a hit. No surprise after discovering the man she'd trusted had suspected her of being a bottom-dwelling traitor. He hated this life that left him suspicious of everyone.

She crossed her arms, protecting her pride from further damage. "If I'd plans to stab you in the back, I'd have already told Kagona where to find Taz."

"I know that now. I didn't then."

"What changed your mind about me?"

"It took time. You're a damn disarming woman." He shifted close enough her folded arms pressed against him. "And no, it wasn't sex that convinced me. Although it was a nice benefit to this mission." Before she could protest, he planted a finger across her lips. "Sipho trusts you, and that counts for a lot. Not to mention what you went through at the Lubbes' to save him. But your actions could've been done to lower my guard. Everything changed once we got to Brugman's. You would've known about his traps, and Kagona wouldn't have risked you with the crocs. Animal actions can be unpredictable."

"Sounds like a witch trial." She jabbed at his chest. "If I had been eaten alive, then I would have been innocent. If I lived, I was guilty. I survived, Ian."

"Barely." The emptiness from those nightmare minutes rushed back. "Mike thought you were dead for a minute. Scared the shit out of me. Logic aside, I'd fallen for you by then." Unable to keep his hands from her, he lightly stroked the sides of her neck and settled his hands on her shoulders. "I wanted to tell you all this last night, but didn't want to ruin the night for either of us."

She let her arms fall free. "Now who's not trusting who?"

"It's a hard thing for me. I've been trained to be paranoid."

"That leaves the question, why would someone working on Taz be in league with Kagona?"

He kissed her forehead and then released her. "My guess is for money or a political agenda." He hoisted the pack over his shoulders. "In these games, there is always something we don't know about one another."

"I can't imagine anyone in the project office capable of committing treason. But I have to trust someone there."

"You have your doubts about Lemmon, but he's my choice."

"Mine, too, although I may regret the decision. I'll trust him for now, but plan to apply your paranoia logic principle."

He chuckled, pleased neither her brains nor her humor had been injured by events.

A whimpering drifted down from the loft above. Ian left Joni standing in the dark to contemplate their conversation. In seconds, he made it up the ladder.

His penlight showed Sipho sleeping. Nightmares. Bloody

hell, what had Kagona done to him? He hadn't spoken on the trip from Victoria Falls, and Ian had cuddled him till his eyes shut.

A double check ensured Sipho was warm. The jacket Mrs. Lubbe had tucked in his pack wrapped around him and dropped well past his hips. His skinny legs were curled up near his chest, and the favorite sneakers covering his feet were snugly laced.

Ian clicked off his light and sat on the hay. Exhaustion plagued his mind and body. He rested his elbows on his knees. His head pulsed. No matter how hard he rubbed, the pressure remained.

Something wasn't right. Why hadn't Kagona personally remained at Sipho's side during the raid on the lodge? Ian had drawn Brugman out of the security room. Why didn't Kagona step in to replace him?

Because he feared facing me. Ian shook his muddled head. *Stuff the ego and focus.*

So what was Kagona's backup plan if he lost Sipho?

Ian shone the light back on the sleeping boy. The light stopped on Sipho's sneakers. Laces. *Damn.* Nomathemba never could keep his shoes tied or even on his feet. Laces made it too hard for him to take them on and off, and Sipho had yet to learn how to tie a bow.

Yet a perfect bow stared back at Ian. Bloody hell. Kagona had messed with his shoes. The laces ensured they wouldn't fall off and be lost.

He untied the sneakers and slipped them off. Sipho stirred.

"I need to borrow your shoes, bud. Think about waking up. Gather your pack. It's time to go see Miss Joni's surprise."

Sipho curled into a tighter ball. The action didn't bode well for moving fast.

Ian tied the laces together and hooked the shoes over a nail in the loft. He slipped the school pack over a shoulder and hoisted Sipho into his arms. Gently, he planted him near the top of the ladder.

"I know it's hard for superheroes to wake up, but Miss Joni is in need of rescuing. We have to get her headed home before the sun rises. Otherwise, her special surprise may be discovered by the bad guys."

Sipho hooked an arm around Ian's neck. His body shook. Yeah, Kagona made Ian feel that way, too. *Bastard.*

"I don't want anyone ever catching you or Miss Joni again." He squeezed Sipho tight. "Hold on. Miss Joni is waiting. You and her and Taz will be going together somewhere far away and safe."

Once down, Sipho switched to Joni's waiting arms. She knelt with him and he clung to her, not saying a word.

Joni straightened as Ian approached. Sipho shifted to hug her tight around the waist. Ian leaned close, encompassing them both, and whispered in her ear, "We have to go. Kagona knows where we are."

"How close is he?"

"I'm not sure yet. Do you have that map I made?"

"Memorized and destroyed."

"Perfect." He led her and Sipho to a side door of the barn and had them wait. He slipped outside and made prearranged birdcalls to his two mates. An answer came from only one direction. *Damn.*

Ian returned and gave Sipho another big hug. "Thabo will take you and Miss Joni to Taz. I'm going to check on Zamani. Make sure Joni steps quietly."

Sipho nodded, only half-awake.

Ian looked up at Joni. "I want you off the ground the minute there's enough light. Stay close." They headed out into the cool night with Sipho holding his hand and Joni on his heels. They hadn't traveled far when they rendezvoused with his mate.

Thabo swept Sipho onto his back. The two had known each other for years.

Ian pulled Joni in close. "Sipho is in your hands. Take care of him for me."

"What if I need to contact you?"

"Lemmon will figure a way. Always has."

"Be careful." Did her voice waver?

"I could say the same to you." His insides knotted, knowing a part of his life was about to walk...to fly away. "Come back sometime...to me."

"Is that what you really want?"

"I'm going to dream about it every night." With one last kiss, he made it quite clear just how honestly he'd answered. Pulling away from her lips nearly killed him. He stepped back with her hand in his.

The longing to hold on and not let go swelled. Instead, he squeezed her hand and released it.

Thabo urged them away, and Ian fled in the opposite direction, knowing her safety counted on his actions in the next few minutes.

Zamani's absence meant he'd been captured or was afraid to signal and be discovered. Either way, Kagona closed in.

The first hints of sunrise lightened the sky as Joni's small group reached Taz. Every step on the trip here had been tainted by emotion. Her feelings for a man who lived without a home in an unwelcoming country were a mess. His lack of initial trust still stung, but the night in his arms told a story of need and loneliness. Should she hate Ian or love him?

The jumble of brush at the clearing's edge combined with the netting had sufficiently camouflaged the aircraft. She placed an affectionate hand on Sipho's shoulder. "Are you ready to meet Taz?"

Sipho nodded, his hollow eyes showing the strain of the last few days. Thabo started pulling brush and the camo net free.

"Then let's get Taz uncovered and ready to fly."

Doubt washed over Sipho's face, but he joined in, helping to fold and drag the net to the bushes. The three pushed and rolled the craft into the open, where Joni could make a short and clean takeoff.

Overall Taz appeared in flying condition. No lion cuddled in the pilot seat or mambas hung from the rotors.

"Inside Taz is the safest place to be." With surprisingly little objection, Sipho trusted her enough to climb into the pilot's seat.

She patted him on the leg and then grabbed a fuel tester. "Don't touch anything. I'm going to make sure a *zorilla* isn't hiding in the engine."

A hint of smile tweaked his lips.

No creatures nested in the air intake, and no damage was apparent from Taz hiding for two days. The early sun painted the field a rich gold with hints of new greens starting to intrude. A few sentinel trees still held morning shadows. Zimbabwe was a beautiful country, where she'd both gained friends and lost confidence in her fellow man. The possible betrayal by a friend or coworker left her uneasy.

She settled both herself and Sipho in the seat and fastened the harness. Her next test lay ahead. Could she fly with a child in her lap? While those at the project office considered he'd be small enough to squeeze behind the seat, none of them considered the fragile state of the boy's mind. Without reassurance, his actions would be unpredictable.

"Ready?" She gave Sipho a little squeeze.

A gunshot sounded in the distance.

They both started. Sipho looked up at her, fear on his face.

"It'll be okay." Her mind argued the case as concern left her feeling empty. How close was Kagona?

She signaled Thabo. "Get out of here."

In a rush, he closed the cockpit as the first morning rays hit Taz. He ran to the edge of the brush to watch takeoff. He wouldn't leave until Taz did.

Sipho still had his head turned in the direction of the gunshot.

"Ian's okay. He's distracting anyone who might be around so we can get away." If only she believed that. Ian's best defense in this case was silence, particularly when outnumbered...unless he had to protect a friend. Gunshots would bring attention to this entire area. That shot had likely been taken at him.

Sipho shook as she locked down the cockpit. "Hold on tight when we take off. It's going to be a fun ride." Her voice wavered as her heart ached. *Focus.*

The push prop behind them softly buzzed to life, allowing the prerotator to engage and spin up the rotor for a jump start. The big overhead blade swept steadily around, much like winding a rubber band on a toy plane before releasing the propeller to spin swiftly.

While Taz was relatively quiet when dropping in for a landing, nearly unseen on radar, and meant to be used at times when no one expected its arrival, its biggest vulnerability was small arms fire.

Thabo stood watch until Joni released the stored rotor energy.

Taz leaped off the ground without a roll and for several seconds hovered ten feet above the ground. She added power to the push propeller and obtained enough lift under the little wings to fly forward. It wasn't long before the barn and fields were left behind.

The prop buzzed at full power as Taz streaked toward Botswana. In all likelihood Kagona and his men watched them go. If so, the authorities knew she had taken to the sky. The haunting question was how fast they could act to catch her.

One arm wrapped about Sipho strapped in front of her. His

heart raced under her palm. "You doing okay? See any animals down there?"

The canopy sloped downward to their knees at a forty-five-degree angle, making it easy to see the ground from the pilot's seat. Sipho shielded his eyes and looked out. It didn't take long before he nodded.

The glare reminded her she had sunglasses buried in a pocket. Surprisingly, they were intact.

Sipho tapped on the canopy, pointing out zebras below.

She pretended not to notice. He tugged on her shirt, and she gave him a quick smile before looking ahead. He tugged on her shirt again and pointed down and out the window. Purposefully she gave only a quick glance.

"I have to fly, Sipho. You'll have to tell me what you see."

He let out a delighted squeal and pointed out an Ndebele village below.

"That reminds me. You did some awesome singing at that *pungwe*. Think you could teach me?"

His chin rose in pride and body relaxed. He might yet decide to talk, but it would be on his terms. The next twenty minutes passed silently.

She checked her watch. The flight into Zimbabwe had passed quickly. The minutes flying toward the Botswana border dragged.

At just under ten thousand feet, Taz cruised at 180 miles an hour—swift for a little rotor craft and a new record for Taz. Obergen would celebrate she had proven the specially designed rotor's ability to sustain high speeds, but in doing so she'd thrown test protocol out the window...and with a child onboard.

Crossing into Botswana's airspace was a risk, but not as great a risk as staying in Zimbabwe. Botswana had little or no radar, and pilots heading out of South Africa regularly, if foolishly, cut across a corner of the country to save time.

Since Sipho sat between her legs, she reached around him to the control.

"Not much father." She pointed out the window. "See that river ahead? That's the Shashi. Across that and we're in Botswana. Then we'll turn toward South Africa."

Sipho drummed on the bare-bones frame of the little copter plane. "Why do you call the plane Taz, Miss Joni?"

Relief swelled within—he felt safe with her.

"It's short for Tasmanian devil. It's a wild creature from Tasmania. Do you know where that is?"

He shook his head."

"An island near Australia. The fun part is cartoons show him swirling around so fast he becomes a blur."

"Like Taz." He pointed up to the rotor and settled back. "Is where we go safe?"

"Yes, quite safe. You'll see animals like at home. And there will be nice people there. One your mother knew quite well when she was a girl your age. They will take care of you until Ian or your father comes to get you."

He looked up over his shoulder with those large, questioning eyes wanting to believe.

She should've put a little more positive sound in her voice. "You'll be safe, Sipho, I promise."

Apparently accepting her word, he helped her fold open a sectional map so she could visually double-check their position. Safety and redemption for her previous failure were only a border crossing away.

Yet a threat waited back home.

Lemmon had limited the group that had knowledge about her flight to Zimbabwe. Others on the project simply thought they were sending Taz on its first cross-country test flight. They had no idea of her destination.

Those in the know included Obergen, a pain-in-the-ass, ladder-climbing project manager. The success of the project meant a promotion for him. Bram Kriegler was in on the trip planning. He lived to fly and without Taz would be hard-pressed to find another ongoing rotor-wing test job. That left two others—the Afrikaner ops director and the maintenance chief. Both were great guys with families to support. Before Ian had destroyed her innocence about her coworkers, she'd considered asking the maintenance chief if Sipho could stay with him. He was native South African and his wife British. They and their kids were super friendly.

How much did she really know about any of them? What made one person betray another? Who would be callous enough to put her in Kagona's hands?

Friendly and nice didn't equate to trustworthy, nor did being a pain in the ass. Lemmon without a question would risk her life if it served his purpose. Bram? Jealous she'd moved in on his territory? Hell, no, they had enough flights for both of them to stay busy. Obergen? She had no idea of the true politics of his position. The same went for the ops director and maintenance chief. She couldn't see any of them as traitors.

When she returned, she'd make an effort to learn more intimate details about them all. Until then, she needed to trust someone.

Ian's phone weighed heavily in her hand. She put a call through, predicting the conversation would not go well.

It didn't.

The conversation ended when a vibration rattled Taz seconds before a brown-and-green camouflage jet whooshed past high and to the left. "I've company."

With that parting note, she hung up and grabbed onto Sipho, anticipating what came next. Turbulent air off the fighter's wingtips swirled into their flight path, violently jolting her and Sipho. Taz jerked and dropped, and the controls momentarily became worthless as the flow over Taz's wings was disturbed.

Sipho squawked in her arms and grabbed tight. She

recovered amid Sipho's screams, soothing him and Taz at the same time.

Asshole. If that pilot knew anything about interceptions, he would have come in differently to avoid threatening their existence from his wingtip vortices. That and other thoughts flashed in her mind as instincts set to work analyzing the situation. Not only did the plane have the circular green, yellow, red, and black symbol of the Zimbabwean Air Force, it was a Chinese-made F7 Interceptor.

Unlike Zimbabwe's British-built Hawker Hunters, the F7s had recently been acquired from China. That meant spare parts and munitions. For her and Sipho, it meant fully loaded weapons and trouble.

At this point, the F7 pilot was checking her out. Kagona might not have secured a good look at what she flew, and Taz appeared quite nonthreatening. The Zimbabwean pilot had to make sure she wasn't simply a tourist who hadn't bothered to file a flight plan.

She unlatched her harness, slipped off the black shirt so her pink one showed, then stuffed her hair up under Ian's khaki hat. No reason to advertise she was the redhead who'd filched a military chopper.

If the pilot had wanted to blow Taz out of the sky, he could have done it a long time ago. The F7 had radar capable of acquiring and locking onto Taz and then shooting a missile up their push prop. As the jet barreled ahead and started a turn back, no missiles were visible. Their presence hardly mattered, though. The F7 had a nice-size gun with enough bullets to make Swiss cheese out of Taz in a few bursts.

She powered way back and let Taz slow without making it appear intentional—at least not yet. Even a rusty pilot studied intercepts. He'd come up on her left side to make a positive identification and signal her to play Pied Piper and follow him home.

Fat chance she'd follow *his* ass. The border lay only minutes away, and she planned to cross it.

The F7 used up precious time in turning and slowing to intercept her flight path. Every second moved Taz closer to Botswana.

Joni did a clearing move to see behind her and confirm only one plane had shown up for the party. No wingman. Guess budget cuts were having an impact. "Sipho, get your head down and keep it down. I don't want him to see you."

Sipho ducked as the F7 prepared to come up on their left. She draped the black shirt over him. His body betrayed his fear as it began to shake. She rubbed a reassuring hand on his back.

"We're going to be okay. Your braid is our good-luck charm. Don't let it go."

His fist tightened around the colorful braid he had finished making on the helo flight from Brugman's.

Her heart pounded. She wiped sweaty palms on her pants. Some confidence. She craned her neck around, trying to see over her shoulder.

If this guy hadn't anticipated her drastic cut in power, he'd still be too hot. The F7 swung back and forth, trying hard to slow. He didn't have the capability to match her air speed and she enjoyed watching his attempt.

She waved from the cockpit as the F7 pilot went by. At least he was far enough away not to see her hand shaking. The less threatening she looked, the more likely he was to underestimate her moves. It would only gain her seconds, but at this point every one counted.

Even with the pilot dropping a notch of flaps, the jet still crept ahead of Taz. The F7 pilot rocked his wings, the universal symbol for "follow me." Right. When zebras flew.

"You can sit up now, Sipho. He can't see you in his six."

Sipho poked his head up and looked cautiously around. He shuddered at the jet dead ahead. "What's a six?"

"Pilots call the tail the six position, like on a clock. If you have weapons and are looking at your enemy's six, it's a good place to be."

"Can Taz shoot?"

"No weapons, but trust me, we're still in a good place. That pilot can't see us, so we have what I call wiggle room. Hang on. I'm going to show you a few of Taz's tricks. Things are going to seem scary, but I have everything under control."

She blatantly lied. No one had tried this particular maneuver in Taz. Theoretically, it should work fine. Of course, "theoretically" had killed cocky test pilots in the past. This maneuver could easily over-G Taz and rip off the rotor or render the craft uncontrollable. But she had no choice. She wasn't cocky, only desperate.

The F7 started a left turn, fully expecting her to compliantly follow him back to base or his choice of landing sites.

"Here we go," she warned Sipho. "If you have to hold on, grab for my legs, not the stick or my arms. Got it?"

"Yebo." Sipho squeezed her thighs so tightly she suspected ten little bruises would pop up.

She dropped Taz's nose to a steep angle, turned to the right, and dived for the deck.

"Whoaaaa…" Sipho screwed his eyes shut.

Her feelings exactly. She set a different heading than they had originally followed, which slightly increased their distance to the river. No reason to continue on the same flight path as before and make it easy for the F7 pilot to find them.

"Is he going to shoot?" Fear echoed in Sipho's voice. She gave him a reassuring one-arm hug.

"Not if I can help it." She had no way of knowing when or if the pilot noticed her departure. "If we're lucky, it will take him a minute to discover we're no longer behind him. By then we'll be two hundred feet off the ground and almost to the river."

"But he goes faster than Taz."

"He'll still have to circle around and find our path to the river. Taz is small and hard to see against the ground, especially from another aircraft."

She didn't tell Sipho that the jet had radar. The very thought sounded rather scary. But the radar was likely the older pulse-type ranging technology with limited capability when looking down. The pilot would have lost any radar lock when he moved out in front of them. As long as they stayed close to the deck, uneven ground, trees, and structures produced radar clutter, making it impossible for it to pick out and lock onto Taz. That left the pilot reliant on his eyes, a plus in their favor.

At around three hundred feet above the ground, Joni adjusted power, letting Taz settle another hundred before blasting back to full power and level flight. Trees on the savanna seemed close enough to touch.

"You can open your eyes now."

Tentatively Sipho opened one at a time, squealing at a herd of giraffes ahead. She avoided flying close and giving away their position if the herd scattered.

Dead ahead lay the sandy stretch of the Shashi River, flowing fuller with the advent of the wet season. The pilot would sight Taz when they crossed. Easy pickings of a light craft on a dark surface. Would he shoot them down over the river or in Botswana, or give up and head home?

The pilot would bring them down if he was aware of Sipho's value and threat. She crossed her fingers, hoping no dire kill message had trickled down through the chain of command.

"Here's the river, Miss Joni. We go to Botswana."

Her breathing accelerated and sweat dumped from her pores. They were fully exposed over the murky waters. This was it. Considering the distance away when they had parted from the jet and the time it would take for him to get on her six, those thirty-millimeter bullets from the F7's cannons could rip through Taz at any moment.

The bullets would shred human bodies like tissue paper.

Five seconds passed. Ten. Halfway across. Twenty. She had one arm wrapped around Sipho and one hand on the cyclic.

Something black blew past Taz's canopy. Sipho screamed.

Damn near scared her to death. "It was just a bird."

Thirty seconds. One minute. Taz buzzed right on into Botswana and kept going. Joni had no plans to turn southeast toward South Africa until well away from the Zim border.

Tension in Sipho's small body relaxed. She couldn't allow herself the same confidence. Kagona was a desperate man and might be willing to invade another country's airspace to destroy a lost prize headed into enemy hands. Another minute dribbled past. Maybe, just maybe, she would bring this kid home alive.

"Flying over Botswana, what can you see?" Sipho chanted, dancing the colorful braid on her leg.

Joni wiped a sweaty palm on her shirt. What had Ian said about her being calm and cool when in her element? Hell, if he only knew.

Affectionately, she rubbed the top of Sipho's head, thinking back to the giraffes they had seen. "A herd of *ntundlas* looking skyward at me."

Sipho twisted to look up at her, his face sad and thoughtful. "When we get to South Africa, will you see me, Miss Joni?"

Her heart ached at his innocence and fears. "Whenever I can, Sipho." She took a deep breath. "I promise."

CHAPTER TWENTY-SEVEN

Joni and Sipho stood hand in hand before three people on an animal reserve in South Africa. The group's gaping stares brought home the arrivals' ragged state. Sipho wore dust-covered clothes and no shoes and had a mostly empty school day pack over a shoulder. She stood dressed in baggy black pants and her tattered pink shirt with brownish blood speckles and a patched shoulder. Her face was tinged camo green, and hair had no shape or direction but at least was hidden under a hat.

Sipho squeezed her hand tighter. She squeezed back. She had a sudden urge to sweep him up and head back to Taz. A slow breath calmed her nerves and helped clear her head. His safety lay here, with family, even if that family belonged to Ian, not Sipho.

A woman well into her sixties pointed at Joni's head. "Ian wears a hat like that." Her lips wavered in an emotional smile. Introduced only by a first name, Joni guessed her to be Ian's mother.

"He still does." A smile rose as Joni pictured Ian standing there in his glorious safari hat and veldschoen shoes, like when she'd first landed in Zimbabwe.

Slowly she slipped the hat off her head, further exposing her ragged state. "He asked I leave it with his godson until he comes to get it. Ian claims he's been away from family too long." A lie, but it seemed appropriate to emphasize the importance of taking good care of Sipho.

She placed the hat on Sipho's head. "This guy will be a big help around here. Ian's been training him to be a tracker. Believe me, he's a natural in the wilds."

Ian's uncle, tall and with a similar chin to his nephew, crouched down before Sipho, who hung close to her side. "You can call me Uncle Charles, like Ian does. We could use some strong arms around here. When you're not schooling, you'll be expected to help. Think you can handle that?"

Sipho looked up to Joni for support. She smiled and he slowly nodded.

"My missus here"—he indicated the sturdy woman next to him with a deep tan and strong arms—"keeps a couple of meerkats around the yard. You ever seen one?"

Sipho nodded again and moved a little closer to Charles.

"They eat scorpions and centipedes. Mrs. Taljaard is perfectly happy about that. However, I could use an extra eye to keep those rascals outta trouble. I came home for dinner yesterday and two of the critters were standing on their haunches looking out the front window."

Sipho laughed at the image. Ian had been right in sending Sipho to his family. Both relief and guilt tugged at her conscience. Was she abandoning Ian and Sipho, brushing them out of her mind and leaving their problems for someone else to clean up?

"You can go climb in the jeep, son." Charles pointed toward an open safari vehicle. Whatever he had to say next, he didn't want Sipho overhearing.

Sipho looked up at her, worry etched in those beautiful deep eyes.

She knelt to equalize their heights. "Don't forget to ask for stories about your mother and Ian when they were kids."

Nervous, he played with the fingers on her hand. "Can you stay, Miss Joni?"

"I have to take Taz home."

"Will you help *Ubaba*?"

An ache spread from deep down. Kids had damnable ways to tug at her heart. "Ian said he'd do all he can to help your father."

"Mr. Kagona say he have *Ubaba*. Will he feed him to crocodiles?"

Ian's mother gasped.

Charles rose to his feet. "Lordy, boy, what a thing to say." His wife placed a calming hand on his arm.

Joni wrapped Sipho into a hug. "Ian is working to free your father. He doesn't want Mr. Kagona to ever do that to anyone again."

"You are strong, Miss Joni. You can help Ian."

"I'd like to, but he's a long way away."

"Can't you take Taz?"

No way in hell would Obergen ever let Taz out of his sight after she'd disappeared with it for days. "They won't allow me to, Sipho."

"Then how will you get back to Ian?"

Kids asked the most perceptive questions. How could she explain that going back to Zimbabwe was likely impossible? Brugman, with his tech-savvy lodge, had recorded their session. If Kagona were smart, he'd have her photo plastered at every border crossing, and he'd have ordered the police and CIO to watch the sky for strange aircraft.

"Ian is resourceful and has friends to help," she countered, feeling foolish and maybe a bit outgunned bantering with an eight-year-old.

"Not like you. You flew me away from Mr. Kagona." His faith in her was staggering and misplaced.

"I could be trouble for Ian. I'm not trained. I make too much noise walking in the bush."

"Ian teach you."

Yes, Ian had taught her a lot and opened her eyes to a part of her life she'd let grow cold. Her lame excuses piled up. Not knowing what had happened to Ian had her nerves in knots.

But how could she connect with him again? He might be in Kagona's custody, injured, or dead. Would she ever know what happened back at the barn? Would Lemmon pass along news, truthful details and not ones made for newspaper consumption?

"Ian is a fine teacher. If I can find a way, I'll help him find your father."

"Promise?"

That simple word hung on the tip of her tongue. Was it fair to give misplaced hope?

"I'll do my best. I promise." And like all the promises she'd ever given, she'd try damn hard to carry through.

Sipho hugged her so tight around the neck, Ian's hat dislodged, and off balance, she landed on her backside in the dust. Sipho tumbled across her lap. A few chuckles from her and giggles from him, and they made it back to their feet.

Sipho dug the toy braid out of his pocket. "Here, Miss Joni. You say this good luck. Use it to find *Ubaba*." He placed it in her open palm and folded her fingers over it.

A mountain of failures crashed down upon her. She'd successfully rescued Sipho, only to abandon him to a life without his real family, with little hope of ever seeing his father alive. Nations wanted everything in his mind, and his closest friend and protector was possibly dead or in trouble.

Sipho snatched Ian's hat up and headed off toward the vehicle, accompanied by Charles's wife.

Ian's mother had nothing short of horror written on her face. "Surely I misunderstood what Sipho said about crocodiles?"

Joni shook her head. "He was captured by some pretty bad

men before Ian could set him free. Expect some nightmares."

Charles crossed his arms and looked to her for an explanation. "There's a man back at the house. He didn't tell us much and asked us to stay quiet about the boy. Told us not to tarry here. We learned a long time ago how to keep matters to ourselves. Is what you told the child the truth? Are you and Ian going to help his father?"

"Ian said you knew Sipho's mother, Nomathemba." Joni evaded a direct answer.

"Yes, a wonderful child who grew up with Ian." A bit of nostalgia played in his mother's eyes.

"I'm sorry to tell you she recently died and her husband, Sipho's father, has been jailed. The CIO is holding him, and they want to use Sipho to encourage him to talk. Sipho also knows things people want kept quiet."

Charles's expression had morphed from suspicious to fully concerned. "Sounds like a dangerous situation for him…and if that man at our house is any indication, for us."

"Ian rescued Sipho from the CIO and had no other place he trusted to send him."

"We're living here now because our lives were threatened in Zimbabwe. We'll see he is well taken care of."

Joni nodded her thanks. "You understand that as long as they have the father, the son is in peril."

Deep lines etched his mother's forehead. "Why didn't Ian come? Is he okay?"

"There was only room enough for Sipho. But Ian's fine." Cold emptiness filled her. Since when had weaving tales become so easy? "He went another direction to secure the area so I had time to get airborne. He's staying to see what he can do about Sipho's father."

His mother took Joni's hand and held it between hers. "Miss Bell, I know you promised Sipho, but I'm not sure if you're brave or half-crazy to go back to such danger."

Guilt surfaced over making near-impossible promises. "There is no guarantee I'll be able to return. It depends on other people and resources."

"Something tells me you are made of the same mettle of my son. It will be hard to keep you away." Hopeful, his mother smiled. "If you do go back, tell Ian we miss him…I miss him. Bring him back to us, even if only for a moment. His father asks about him."

"Is he here?"

"No." A heavy sigh escaped her lips. "I hear from him about once a month when he sends money home. He comes back off and on but never stays long. Always asks about Ian. Crazy about his son."

Charles gave his sister-in-law a hug around her shoulders. He looked out across the dirt strip. "The men are almost done refueling your craft."

Joni walked with them to the jeep, where Sipho stood with Charles's wife.

Ian's mother pulled Joni aside while the others loaded up. "I live every day with no guarantee the men in my life will come home. Tell Ian I love him, will you?"

All Joni could manage was an encouraging nod. No more hollow promises. No more lies. In reality, she had no way to infiltrate the country again and no justifiable need to go back to Zimbabwe.

She had completed her mission and flown Sipho out. Now it was time for the qualified spies to move in if Lemmon felt the data in Sipho's head justified an excursion. Ian had friends to help him determine the best way to get Mukono out of jail. A political or legal solution might work better than an escape attempt.

But what if Kagona had Ian or he was injured? Who the heck was his backup? Surely not Lemmon. Could she live day by day without knowing what happened to the man haunting her thoughts?

The open-top vehicle started up with Ian's uncle at the wheel. He looked over his shoulder at Sipho. "Did Ian teach you how to tell the difference between white and black rhino tracks?"

"Yebo." The truck pulled away with Sipho caught up in the conversation and distracted from looking back. Ian had been right about bringing him here.

Joni watched them go, feeling alone and a little lost. Had these last days full of terror, friendship, and sacrifice been for nothing? New memories jammed her head, but her life was destined for fewer fulfillments instead of more. The friends she had back at the project office could be personal threats. Trusting anyone there would be nearly impossible.

The men had finished adding enough fuel to get her back to Hoedspruit and were moving their equipment away. Time to fly home to face Lemmon, put on the tough, innocent face for her fellow project workers, and watch for a knife in the back.

What had become of her life—one with high ideals she fought for, strived for, and lived to achieve? Had failure brought her so low she'd forgotten the only person to make life worth living was herself? If she wanted a fulfilling life, she had to fight for it.

Taz had become an oven in the daytime heat on the dirt strip. She opened the cockpit and hot air blasted her face. Determined to let nothing slow her down, she climbed in and strapped into the harness. Once airborne, she had plans to make. The main focus of her future was not whether she should go back to Zimbabwe, but how.

The landscape beneath had morphed into a vast wilderness of trees, hills, and range belonging to Kruger National Park. She approached Hoedspruit Air Force Base nestled midway along the enormous park, which bordered Mozambique. The runway, built long enough to handle a space shuttle landing, was the base's most prominent feature. Taz required only an inch of that yardstick.

Below, the majority of military hangars and facilities had been strategically camouflaged or built underground to blend in with the natural environment. Most of the visible buildings belonged to the rustic adjoining civilian airport.

The Tasmanian devil, with minimal fuel and clearance to land, swept over the runway threshold, over the target from the flight demo, which seemed an aeon ago, and set down easily on the runway. With the push propeller driving Taz along, Joni taxied toward the project hangar.

The doors slowly slid open. A marshal signaled her to stop outside. She cut power and released her harness. Taz had done his job.

Dreading the confrontations to come, she used the time until the rotor stopped to assess those walking toward her. Her trusted project leader, Obergen, led the pack, followed by Loots from ops and Tinibu from maintenance. Lemmon trailed. The only one missing was Bram.

The recovery team was already tilting open the nose canopy by the time Obergen made it to Taz. She stepped down with mixed feelings about being home on South African soil.

"We've been worried sick about you," Obergen greeted.

More than likely about Taz, but Obergen did sound relieved. He'd probably regretted his decision to assist Lemmon when she didn't immediately bring his equipment home.

Obergen gave her hand a quick shake. "You can't imagine the stress here with only one report, and then after two days suddenly announcing you were two hours out from landing."

"I had to borrow a phone. Mine got confiscated. And I didn't want to use Taz's radio until entering South African airspace."

He looked past her to Taz. "Did you get the material?"

"You mean the boy?"

Obergen showed no surprise at her mention of a child. He'd known all along her goal was to bring Sipho back.

"I was able to get him to safety." She watched his guilty look morph to puzzled.

"I assumed you were bringing him here." Obergen glanced back at a well-dressed black man in a suit who waited far in the background. Perhaps a representative of the government or the South African president? Neither he nor Obergen appeared happy. "Where is he?"

She looked past her boss. "Excuse me, but Mr. Lemmon is waiting for a quick debrief."

Obergen opened his mouth to argue, but others stepped forward to congratulate her. Tinibu, full of friendly grins, patted her good shoulder, thanking her for bringing Taz back in one piece. Loots gave her a mock salute and a handshake.

Lemmon stood way back, looking noncommittal. He, too, noted Taz only had one occupant. Joni wished she'd left the phone in Zimbabwe so Ian could have checked in and given Lemmon news about his welfare. Instead, the phone burned a hole in her pocket.

Questions started flying about the trip and Taz, but she headed toward Lemmon, not quite sure how to say what needed to be said.

Lemmon indicated they should walk together away from the crowd. In few steps, he got right to the point. "Did all go well with securing Mukono's son?"

"He's with Ian's uncle and mother at the animal reserve. Kagona will learn that Sipho didn't show up here. When he does, it's only a matter of time before he finds him."

"I've someone in place to keep an eye on the boy. I'm pleased you were willing to trust me."

Joni snorted. "Not totally and never will. At the moment, you're my best ally. Hopefully I didn't choose poorly."

"A cynical touch. There may be hope for you yet."

"I'm not one of your operatives, although you seem to use me that way at will."

"Considering what you've been through—more than once, I might add—I'd say you're as good as those I've trained, if a tad naive."

"Naive? Not so much that I can't imagine you chose this project not to test out Taz, but because you had good reason to believe there was a traitor in our midst. One with a back door to Zim's chief CIO operative. Surely you must wonder why. Was that what drew you here? Are you somehow monitoring Kagona's communications?"

Lemmon's eyebrows rose. "Well, well. I may have underestimated you again."

"When will you quit lying to me?"

"Lying? No." He shrugged. "Omission of certain information—yes."

Joni stopped to stare through Lemmon's eyes into his wretched soul. She sure as hell wanted to make his conscience burn next time he deceived her.

"You're not to touch Sipho. If you want something from him, tell me and I'll do my best to find out what you need. Kagona had him for hours, and if he witnessed even half of what I did at that place, he'll have nightmares for life. I don't need some psychologist stirring all that up and closing his mind down for good."

"I'm not inhuman, Miss Bell. He may well have timely information, though. This might be a good week for you to take time off. You know to recover from your traumatic flight."

"And spend time with Sipho to drag information out of him."

"Like you said, better you than us."

"You're a bastard, and for once my answer to you is *no*. I'm not one of your operatives. What matters to me is that Taljaard comes out of this mess alive and Sipho has a safe place with family to live. I'm not going to sacrifice all to prove Zimbabwe is making nefarious deals with America's enemies." She reflected on how the next words would go down. "Unless…"

"Yes?" Age lines crinkled around Lemmon's expectant eyes, and it egged her on.

"I want to go back. I need to find out what happened to Taljaard."

"No."

She lowered her voice. "Taljaard was quite willing to discuss the ramifications of what Sipho might know. If you want to discover what's in the boy's head, Taljaard is more likely than anyone to draw it out."

Lemmon tilted his head, seeming to consider the idea. "Not a bad option to consider, however, it creates quite a conundrum. You see, I need Taljaard in Zimbabwe to keep his ears and eyes open. And since you returned with his phone, my communications have taken a hit. It will take time to reestablish contact."

"Briefing in twenty," Obergen called from afar.

She nodded at her boss and then refocused on Lemmon. "I can help do that. Get me into the country to make contact and help us with Sipho's father. In exchange, we'll be your eyes and ears for whatever you need."

"For your own health and safety, I can't let you return. You've never been trained on covert activities. Alone, you'd be an easy target."

"I can do this. Ultimately I won't be alone."

"You have no idea what happened to Taljaard. He could be dead."

"That's what I want to find out. For Sipho's sake, I have to know."

Lemmon shook his head.

Joni had no energy left to argue with Lemmon, but she wasn't giving up. Give him twenty-four hours to mull over her offer. Desperate times changed minds.

She stalked after Taz, which was being pushed into the hangar.

"Miss Bell," Lemmon called after she had put a good distance between them. "I believe you have my phone."

She stopped, reluctant to hand it over, but not sure why. Slowly she dug it out of her pocket. It was the object Ian had used to contact Lemmon over a week ago and start the round-trip journey she'd just completed.

It had served its purpose. She handed it over, feeling another connection to Ian snipped.

Feet pounded the tarmac behind her. A second later, a flight-suited Bram swept her up and spun her around. "*Aweh.* Good show, *bokkie*. You should've seen the way everyone paced around here when we lost contact." He set her down and took a good look at her tattered state. "*Blerrie*, what happened to you? Look like a damn lorry ran you over."

He picked at her tattered and bloody shirt, then rubbed a finger across her face, still stained with camo grease. "Drinks on me after debrief. No kidding, this is going to be an amazing story." Bram rubbed his hands together, bolstered by the thought of new rumors for the pilot mill.

She grimaced in pain when he gave her shoulder an excited whack. Of all those who had welcomed her home, at least he'd noticed she hadn't survived unscathed.

Those very thoughts left her insecure. Which of these well-wishers had betrayed her mission to Kagona? Lemmon crossed his arms and looked perturbed.

"Come on, Bram, I'll give you a few teasers about my last two days before the debrief."

He crunched another hug around her shoulders.

"We'll speak again after debrief, Miss Bell." Lemmon took a few steps away, then turned back. "By the way, did you teach this child to eat worms, too?" He walked away, not waiting for an answer.

"What was that all about?" Bram asked.

A strange hope flushed her body. No one knew what she'd done with the kid to survive in Colombia. Could he still be alive? Had someone talked to the boy after her rescue...his rescue? Of course, eating worms was a typical survival trick. Maybe giving her hope the boy had survived was Lemmon's sick idea of a reward for a successful mission. Would he ever tell her for sure?

Lemmon failed to show at debrief. She kept her story of the trip to Zimbabwe short, focused on the aircraft, and ignored events outside of Taz. She used the excuse that Lemmon had requested, for the safety of those she encountered, to withhold sharing details.

After the brief, she collected her car keys from her desk and tossed them into a day pack. Home lay a twenty-minute drive away.

Bram corralled her in the hallway. "Come on, Joni. Give me more tidbits. I can tell you had a roaring adventure and I'm swimming in jealousy."

"Another day, Bram. I've barely slept."

"Go home. Take a nap. Then come on. I got a supply of Castle lager at the apartment."

She slumped back against the wall, thinking about all the times they'd shared a drink yet never built a rapport beyond work. Ian had set the bar high for anyone to ever fill his shoes.

"I'm taking a week off. But to keep you from dying with curiosity, I'll give you a few hints. The dirt and blood came from a run-in with a chopper." She grinned. "I won."

His eyes grew joyfully round. "No shit."

"The camo." She slipped an arm around his and paraded him toward the exit. "Did you know crocs will crawl onshore for dinner?"

A shadow materialized ahead. Lemmon.

Bram didn't need to be told. He walked past Lemmon and headed down another corridor.

Joni crossed her arms, a motion becoming common every time Lemmon graced her with his presence. "I doubt there is much I can add to what you already know. And if your little hint back there was meant to motivate me, it would indicate you've deceived me once again."

Lemmon leaned against the wall opposite her. "You know about my business—it's secretive. Company and country politics take certain, ah, choices out of my hands."

"What are you saying?"

"Frequently operatives are not fully aware of the results of their missions."

"I was a military pilot being used by the CIA, not a damned operative."

"Yes, I had that pointed out to me by several of my colleagues." Lemmon stared down at his shoes before looking her in the eye. "At the time, they were quite concerned about your knowledge of the Colombian boy and his family."

"Are you saying the defense minister's son is still alive?"

"I'm saying nothing. Couldn't even if I wanted to. You, however, can make any assumption you want."

She let the impact of Lemmon's innuendo sink in. A second later, rage at the heartbreak and guilt she had suffered exploded. "You left me hanging for two years? Two long years. You son of a bitch." She headed for the door.

Lemmon raced to catch up. He knew better than to touch her, so he made a calculated leap into her path. "You should at least consider the facts, which you seem to have so cavalierly thrown away."

Joni stopped. "It's inhuman to withhold something like that. I wasn't just *anyone*. I spent a traumatic month with him. He mattered to me."

"I am a loyal employee of a very complicated and protective

organization." Lemmon spread his hands. "*I* kept nothing from you. And the last time I saw you, if I remember correctly, you were surrounded by medics on the way to a hospital, fever eating you alive. You babbled about things that had happened. A bad situation, really. I was quite concerned about your health."

"My health? Not likely. You had a chance to discuss this with me before I left with Taz. Why didn't you?"

"Your reaction at seeing me here in South Africa told me the less we talked, the better. If you had known about Sipho, you wouldn't have gone, regardless of anything I would have revealed beforehand. You have a propensity not to believe me. I knew, though, once you arrived in Zimbabwe, you wouldn't be able to leave without the boy. Your personality profile says you will try harder to succeed if you have failed before. I made a correct assumption, wouldn't you say?"

"Do you have to justify everything so you can sleep at night? Who would have taken Sipho if he had landed with me?"

"Our diplomatic ties with the South Africans are delicate."

"Can't you ever tell the truth?" She shoved open the door to the parking lot and stepped outside. Lemmon followed.

"I don't lie to you, Joni."

"You don't need to. You have manipulation down to a fine art. I still haven't come to terms with a traitor in the project office. It could be you."

"Yet you chose to trust me."

"I had to pick someone. I don't let personal distaste stand in the way of objectivity."

"Wise choice. I received calls while you debriefed. It seems money is being distributed from an account we've been watching."

"Chitima?"

"You know?"

"Taljaard told me. But it wasn't the only time I heard the name."

Lemmon practically drooled with anticipation.

She took a healthy pause before answering, enjoying every second. "Before we landed in South Africa, Sipho asked me why Mr. Kagona wanted to know about the train. I wasn't sure what he meant, until Sipho mentioned that *Chitima* is the Shona word for train. Sounds like Kagona was pumping him for information on what his father knew about the account."

"Does seem that way."

"So what do you think is happening? Are the illustrious leaders dividing the spoils?"

"This account is different than most of the personal ones we've been following." Lemmon stuffed his hands in his pants pockets. "It's being used more like payoffs for transactions or bribes. We don't have exact details yet, but at least some money is going out of the country."

"To island banks?"

"No. Mozambique and China."

"Who is there?"

"More like what. One of the routes Zimbabwe uses to connect to the sea is via Beira, a port in Mozambique. Some of the money went to a Chinese shipping company. They own a good number of the ships bringing in oil and cargo to Zimbabwe."

"I get the gist. Why is that important? Wouldn't that be typical if they're buying oil or shipping in goods?"

"Like I said, nothing about this account seems normal. We believe this time something is about to be shipped out from Zimbabwe."

"And you think time is running out for you to discover what and to whom?" She offered no sympathy but saw opportunity.

"We need knowledge and proof, or it will be too late. I'm sending some people in, but knowing where to start or what to look for is the hard part."

"Sounds like you require Taljaard's help."

"You could tell us where to look for him."

"I'm not even sure he's still alive. There was a gunshot nearby when we took off. Taljaard was drawing Kagona's men away."

"Assuming he is alive, where would I find him?"

"He could be two feet away from one of your people, but he knows there's a traitor in our midst—he wouldn't trust them enough to make contact."

Lemmon, likely catching emotions streaming across her face, smiled ever so subtly. "I can only guess from your shattered appearance, and yet survival, that you made an impact on him. You asked to go back. You leave in forty-eight hours."

"This is what you had in mind all along, isn't it? You know my answer. Only this time I want to be prepared. I'd like a crash course on tricks of the trade and someone to check out the damn slice and bruise on my shoulders. I'll need some communication devices, any special equipment you think useful, and I need to replenish things Ian might require to pull off this job. Plus, a thorough vetting on Mukono and the leaders in Zimbabwe. I want to know who runs Kagona and his weaknesses."

"Such a short list. Taljaard could fill you in on much of this."

And she'd love to coax the information from him but doubted they'd have the opportunity for play. "I've learned not to count on others in this slimy business. My participation comes with a full brief, not one filled with lies."

"You're building quite a chip on those shoulders, Miss Bell. Any other demands?"

"Obergen is not likely to let Taz out of his sight. How do you suggest I get into the country?"

"Leave the transportation details to me. Get some rest. We'll get started in the morning. I'll warn you, though. With this being a complicated field op, you're not going unless you pass a test of my creation. Training might take longer than forty-eight hours."

Lemmon stepped away and headed down the sidewalk. Joni cleared her throat.

He stopped and faced her. "What is it, Miss Bell?"

She signaled him closer and waited until he stood close once again. "I have a few numbers to add to your Chitima file."

Lemmon's eyes lit up, and a sly smile wrinkled one cheek.

"Don't get too excited. My memory isn't like Sipho's."

"I'm not complaining." But he had dug out his phone and hit the recorder.

She carefully repeated the numbers that had been going through her head since Sipho had helped her remember them. Her insides clutched at the memory of his laughter as she messed up the first several tries to remember the long sequence.

Lemmon walked toward a waiting vehicle with tinted windows. He talked through a lowered window with someone inside. The South African president's man? A minute later the vehicle drove away, leaving Lemmon behind.

A tan SUV pulled out of a parking spot and stopped next to Lemmon. He climbed in the passenger side and the driver sped off. Obviously Lemmon had a team in South Africa.

She leaned back against her little car and dug into her pockets. Her fingers tightened around the square braid Sipho had given her. She brought it out and rolled it around on her palm. Could she keep his father alive—or, for that matter, Ian?

Exhausted, her mind still spun with unanswered questions, personal doubts, and uncertainties as to whether Lemmon sending her back hadn't been planned from the start. Had he hoped she'd seduce Ian, coax him to further some CIA agenda?

Why had the world suddenly become so opaque? Her vision of Ian's arms warming and holding her became hard to see.

Joni opened the car door. At her feet, something light stood out against the dark asphalt. She scooped up a piece of bent hay. It must have fallen from her pocket.

She rolled it between her fingers. To hell with any

misgivings. One constant remained. Ian had filled up a part of her she hadn't known was empty. No way was she waiting a lifetime for him to track her down. Next time she wrapped her arms around him, she wasn't letting go.

She tossed the hay onto the dash and climbed in the car. After the motor buzzed to life, she shifted into gear and tore down the empty stretch of base road. An odd sense of freedom lightened her soul. Tomorrow couldn't come soon enough.

"You won't have long to wait for help, Ian." She paused to smile. "I want my panties back."

Thank you for taking the time to read UNDER THE RADAR. *If you enjoyed the story, the greatest way to say thank you to an author and encourage them to write more in the series is to tell your friends and consider writing a review at any one of the major retailers or on Goodreads. It is greatly appreciated.*

BOOKS AND NEWSLETTER

Authors love to get feedback, so stop by Sandy's website, or blog, True Airspeed, and make contact with her.

Website: www.sandyparksauthor.com
Blog: sandyparks.wordpress.com

If you would like to be informed of upcoming books, please sign up for Sandy's newsletter via the website. It's sent infrequently and will be fun as well as informative. Also look for future additions to the website, Pinterest, or blog with photos and information on locations, settings, or aircraft mentioned in the Taking Risks series.

Join me:
Facebook: Sandy Parks-Writer
Twitter: @SParksauthor
Pinterest: parks3353

Thanks for your support, and stay tuned. More books are coming.

To see what happens next with Joni and Ian, read the attached first chapter from *Off the Chart*, book two in the Taking Risks series, available in ebook or print.

If you enjoyed Sandy's writing, you might like to try her romantic thrillers in the Hawker, Incorporated series which starts with the multiple award-winning book *Repossessed*.

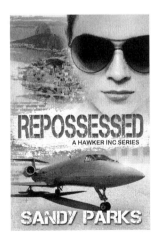

AUTHOR'S NOTE AND ACKNOWLEDGMENTS

Take a country with US State Department travel warnings, add in a sprinkle of foreign politics, a strong-willed woman, a sexy former Special Forces soldier turned wildlife activist, and mix with new aviation technology to provide fodder for suspense.

This book idea came from my husband's visit to Zimbabwe years ago on a flight test mission and my desire to write about women test pilots, who are just as smart, courageous, and sexy as their male counterparts.

Flying isn't just a hobby in my family—it's a passion. My father took me up in a small plane at a young age, and after that I always knew I'd become a pilot. After graduate school, while my husband specialized in the daring and dangerous job of test pilot at Edwards Air Force Base, I worked on a flight test project team and learned to fly light airplanes.

I wish to thank—and hold blameless for my mistakes—those whose expertise I sought to add plausibility to this story.

Retired USAF Pararescue Master Sergeant Dave Young
Former USAF Pararescue Jarrod Honrada
Pilot Ian Taylor with many flying hours logged in South Africa
Helicopter pilot Elena DePree
Pilot and husband Scott Parks

Friend and pilot Bill Weiler
And British friends Sally and Sarah.

I also took a few liberties, for simplicity's sake, to make the story easier to follow for those not as familiar with the locations. For example, the current helipad in Victoria Falls is near the Elephant Hills Hotel instead of the Victoria Falls Hotel.

Excerpt from

OFF
THE
CHART

A TAKING RISKS NOVEL

SANDY PARKS

CHAPTER ONE

Africa

Mobile. Tactical. Tiny.

Joni Bell manipulated a drone, no bigger than a dinner plate, over the African bushveld toward tents of a makeshift camp. Bushwillow leaves ruffled in a breeze that masked the slight buzz of the eye in the sky but did nothing to halt the drips of sweat running toward her eyes.

Doubt inched into her mind. She had one chance. *Focus.*

Four men in fatigues with enough weapons to remove her as a problem came into view on her phone attached to the drone's command console and controller. She guided the craft past an all-terrain vehicle and over the broad spread of a knob thorn tree, where the men huddled. One knelt and motioned at the screen of an electronic tablet. She zoomed the lens to see they studied a satellite view. Of what the limit of her zoom and moving leaves and twigs of the tree made it unclear.

Her minimal covert field training had the potential to kill her, but one man, one who offered hope for a fulfilling future, believed in her natural skills. *Prove him right.* Her success and his life depended on it.

A civilian, dressed in the light green shirt and olive khakis often worn by professional trackers, approached the others. One more advantage for the enemy. The kneeling man rose and addressed him. They hunched over the tablet, discussing something on it.

A tap on her phone directed the hovering drone past the protective edge of the tree canopy. Her focus zeroed in on the tablet display.

A low-battery warning flashed. She tapped it off. Another second and she'd learn the exact target.

The tracker pointed at tangled brush on this part of the savanna. The leader shifted. His arm covered the tablet. The tracker swept his arm around to indicate something behind them. The other men looked where he indicated. One soldier glanced up. He yelled and pointed skyward, directly at the camera.

Crap.

No time left to determine the actual target, but a prominent landmark had stood out on the leader's tablet—an isolated runway with a unique configuration.

Joni whisked the drone off in a nebulous direction from her position. She disconnected her phone but left the console on and stashed it out of sight. She grabbed her day pack and hustled to create distance from the men.

They had a vehicle.

She was on foot. But they had to figure out in which 360-degree direction she had escaped in a large, wild countryside.

Stones lined a dry gully ahead. The V cut provided cover and aimed away from the dirt road the men used. Minutes passed, and with each one her chance of successful evasion increased.

The rattle of a fast-moving vehicle over rough road sounded in the distance. Were they following the drone? If so, it wouldn't be long until it ran out of power. Before then, the auto-land feature should cut in. She'd disabled the auto return so it wouldn't come back to its starting GPS and pinpoint her location. Hours spent flying recreational drones with her nephew had helped, but one major difference remained. This time she played for people's lives.

Her heavy breathing, mixed with the constant chatter of grasshoppers, made it impossible to tell if the vehicle moved closer. She scrambled over slabs of grayish rock and up the red clay sides of the gully.

Sweat dribbled down her back. Dry, wild grass crunched under her feet. She headed toward softer soil only to see the hardened impression of a paw. A big one.

Rain last night had softened the ground while the night hunters prowled. Now the day's sun was well on the way to hardening their tracks. She dug out her meager handgun and shoved it into her belt behind her back. Best to be prepared for the nonhuman dangers as well.

She hid behind a copse of mopane brush and stopped to listen. The damn chirps of birds in nearby grass masked other sounds. Scare the birds and they'd swarm, alerting the men. In the back of her mind, a warning rose about cheetahs hunting during the day and the birdlike chirping communications they made. Great. Did they eat people?

The air seemed oddly quiet of man-made noise. The truck had stopped. Where?

With soft steps, she set out, only to hear an engine start up. They had paused to track and were moving again...in her direction.

The engine strained as the driver maneuvered over uneven terrain, but the 4x4 progressed steadily. She headed behind and up a hill in an area difficult for vehicles to traverse. At her next pause, the vehicle noise had silenced. Likely she had forced them to proceed on foot.

That gained her time, but not much. The moves she made in the next minutes held the key to survival. She longed for Ian

Taljaard at her side. His wisdom of the African wild could guide her steps, give her strength. How would he slip away from these men?

His answer whispered in her head. *Become one with the land.*

Yeah, right. Easy for Ian, wildlife rancher and former British SAS. The African bush was his playground. He'd probably had a childhood pet that sported a mane and roared. What she wanted was a helicopter to fly out of trouble.

Not far away, the grasses became heavily dotted with trees and entangling brush. That added cover would work to her advantage. She tapped out a quick text message on her phone, highlighting her discovery. A least the message made it out, even if she didn't.

She took a direct path toward the thicker foliage on the flats…and so did the men. They made little attempt to hide their movement as they pressed closer. Cocky bastards. Her hands shook. No help would be coming. She alone had to succeed.

Joni's feet scrambled in conjunction with her mind. Step quick and careful. She had no way to outrun these men, and turning to fight with their weaponry shouted lousy odds. Still, she was relatively sure they had yet to see her. Her chance of escape rested on whether or not the tracker had stayed with them.

Past a large tree ahead, the ground grew rough. She leaped over a small mound into a huge rut. Soft soil shifted under her feet. A stench overwhelmed her.

Shit. Literally. She'd landed in a rhino midden, a communal dung heap. Shade from the tree kept piles of dung soft, and from the pungent odor of the piles—fresh.

A quick check revealed the horned, tanklike animals must be scouring a different pasture. She lifted a foot to climb out, but stopped. Ian had taught her the best places to hide were where no one wanted to go.

A dung beetle rolled a compacted prize many times its size

out of the long rut. The piles of poop in and around the depression offered opportunity if she chose to take it. Her insides threatened to retch.

In an eerily calm moment of clarity and with no time to waste, she yanked a handkerchief out of her pack and propped herself against the short back wall of the rutted depression. Cool moisture of the mound seeped through her pants. No time for second-guessing her decision.

Joni covered her legs with intact chunks of grass-ridden scat. Her mind, in a strange attempt to avoid thinking about the odors, classified the poop as from grazing white rhinos. Perhaps Mr. Lemmon had been right about her needing a vacation. She was losing it, becoming certifiable. Likely headed for downtime in a locked ward.

Dry ground litter crunched under boots in the direction of the big tree. She put the cloth over her face, folded one arm over it, and then loaded on pellets and muck with the other arm. Once done, she worked her free hand and arm into the slimier mass. At any moment, the men should pass her position.

Although muffled, sounds of the living bushveld came through the load of dung. The men had gone silent. No footsteps, no voices, no weapons jostling...just dang crickets chirping away. Breath became hard to draw. Panic welled, and she fought it down.

Then a faint sound reached her ears. Not sharp or heavy enough to be from firearms. Her gut clenched on recognition of the electronic clicks of a digital camera and then all-out laughter.

Defeated, she flung down the dung-slick arm covering her face and lifted off the handkerchief. One of the men leaned over the midden. To his credit, he didn't laugh. Instead, he straightened and pointed into the air. Above them floated a dull bluish-gray drone, bigger and meaner looking than the one she'd used.

"You're not the only one who plays with toys, Bell."

The humiliating failure left her strength gone, along with hope to rescue Ian. Her throat swelled closed. Each rancid, disgusting breath of air took effort.

No one would ever know if Ian had survived the last mission or sat inside the hellish confines of a Zimbabwean intelligence cell. She'd lost her opportunity to find out.

"No second chances, I assume?"

The soldier shook his head.

"Any possibility I passed?"

"I merely offer a report." He chuckled and pointed at the camera. "Mr. Lemmon is the final judge."

Photo by Robert Vanelli,
Exposure Photographic Art Studio

Sandy's romantic thrillers are award winning. *Repossessed* collected a 2013 Kiss of Death Daphne du Maurier Mainstream Mystery/Suspense Award, a 2013 Maggie Award for Novel with Strong Romantic Element, and a HOLT Award of Merit.

Flying and science are evident themes in Sandy's manuscripts. She is a hydrogeologist by training, with an MS in geological sciences, and has completed additional graduate engineering course work. She has taught at a university, worked on a military project for the Air Force Flight Test Center, worked as a design engineer for a civil engineering firm, and done computer modeling and field studies as a hydrogeologic consultant. Keeping up with her two sons, she also learned how to fly and dive and survived sparring for fifteen years as a black belt in kenpo karate. She has studied in England and Italy, traveled to South Africa, Egypt, Asia, and South America, and still travels to places of interest all over the world so she can make her stories richer.

54818546R00205

Made in the USA
Columbia, SC
06 April 2019